NAVIGATING BIODIVERSITY

First published in 2024
by Riverside Press
an imprint of
UniPress Books Ltd
World's End Studios
London SW10 0RJ
United Kingdom

Copyright in the Work © Unipress Books Ltd 2024
Artwork copyright © Robert Fiszer 2024

ISBN 978-1-7397988-9-5
E-book ISBN 978-1-917226-00-4

All Rights Reserved.
No part of this publication may be reproduced,
stored in a retrieval system or transmitted in any
form or by any means, without prior permission
in writing from the publishers.

British Library Cataloguing-in-Publication Data
A catalogue record for this book is available
from the British Library.

Publisher: Jason Hook
Project manager: Katie Crous
Design: Alexandre Coco

10 9 8 7 6 5 4 3 2 1

Printed in China

unipressbooks.com

FIND YOUR WAY THROUGH BIG IDEAS

RICHARD KEMENY

RIVERSIDE PRESS

INTRODUCTION	6

1 ORIGINS OF DIVERSITY   8

How did the platypus get its bill?	14
How did the herring teach the smelt not to freeze?	16
How do we tell a giraffe from a tree?	18
Is there strength in numbers?	20
How much is a coral reef worth?	22
Do mountains make life more interesting?	24
What makes a biodiversity hotspot?	26

2 ECOSYSTEMS   28

When does a river become an ocean?	34
What happens when a whale dies?	36
What links a plant with a volcano?	38
Is a shark more important than a scallop?	40
How does the butterfly deal with competition?	42
Would you want to hitch a ride on a shark?	44
Why haven't sea turtles taken over the world?	46

3 THE UNSEEN WORLD   48

Should I shower every day?	54
What makes snow turn red?	56
Could we survive another 'Dust Bowl'?	58
How do plants talk to each other?	60
How many microbes are in the ocean?	62
Are viruses a good thing?	64

4 OPPORTUNITIES   66

Why should I care about mangroves?	72
What fuels our survival?	74
What have wasps ever done for us?	76
How do other species protect my health?	78
How do we get aspirin from a tree?	80
How did the tourist save the gorilla?	82

5 TECHNOLOGY 84

Can you spot an elephant from space?	90
How do birds sleep when they fly?	92
If a tree falls in the woods, is the sound useful to scientists?	94
If a fish committed murder, could we tell which one did it?	96
When is a jellyfish no longer a jellyfish?	98
Will AI save the world's biodiversity?	100

6 THE ANTHROPOCENE 102

Where can I find wilderness?	108
Are cities good or bad for biodiversity?	110
How did the chicken take over the world?	112
What if humans never invented sailing or flight?	114
Am I eating plastic right now?	116
Were there once lions in Europe?	118

7 LIFE UNDER THREAT 120

Do young Florida panthers move far from their parents?	126
Why do people want to traffic pangolins?	128
Can corals keep up with climate change?	130
Why are Mexicans making bricks out of algae?	132
Where is the lionfish going?	134

8 CONSERVATION 136

How do we know when a species has been saved?	142
Can Noah's seed ark save us?	144
How many bison does it take to rebuild an ecosystem?	146
Should we gene edit mosquitos out of existence?	148
Can a river have rights?	150
What role can indigenous peoples play in conservation?	152
What can I do to protect biodiversity?	154

FURTHER EXPLORATION	156
NOTES ON CONTRIBUTORS	157
INDEX	158
ACKNOWLEDGEMENTS	160

INTRODUC

Biodiversity is the variety of life – from the smallest of microbes to the largest organism on the planet and everything in between. As our understanding of it has deepened, so too has our realisation that we are dependent on it to survive. The natural world sustains us, bringing a wealth of benefits. Biodiversity, in all its complexity, provides us with a deep-rooted support system that has allowed our civilisation to prosper and thrive, from the water we drink, to the air we breathe, to the soil that feeds us. Some of these benefits can be valued in economic terms, others cannot.

Is the success of *Homo sapiens* down to human intelligence and ingenuity, or is it really down to microbes? The world is filled with microscopic organisms too small to be seen with the naked eye, but who live all around us – and inside us. They have transformed our planet, and perhaps our own evolution, in ways that we are only just beginning to understand.

Alongside great strides in ecological research, we have created new technologies to help us monitor, understand and conserve biodiversity. We can follow sharks as they dive into the deep ocean and accompany migratory birds on their flights around the planet, learning more about their unique behaviours and ecological roles.

Our knowledge has grown in tandem with an awareness that our use of the planet has come at the severe detriment of other life forms. We have scarred our lands in search of minerals, uprooted billions of trees, stained our water and soil with pollution and altered the atmosphere with fossil fuels. We have changed the very functioning of our planet, the consequences of which are becoming increasingly clear.

In our expansion around the world, we've taken a few select species of plants and animals along with us. This has been a boon for us and them, but at what cost? Are we making the planet blander by spreading certain species around? Is our interaction with, and as part of, biodiversity creating new life that may never have existed? What does human expansion mean for the diversity of animal, plant, fungal and microbial life that wants to live here too? Now we have the tools to alter evolution, potentially remove entire species, or even bring back others from extinction, should we use them?

By exploring some of the questions at the heart of today's debates around biodiversity, and their scientific and philosophical underpinnings, this book will give you a deeper understanding of the variety of life that exists on our planet, how humans have changed it and how we might best preserve and restore it for future generations.

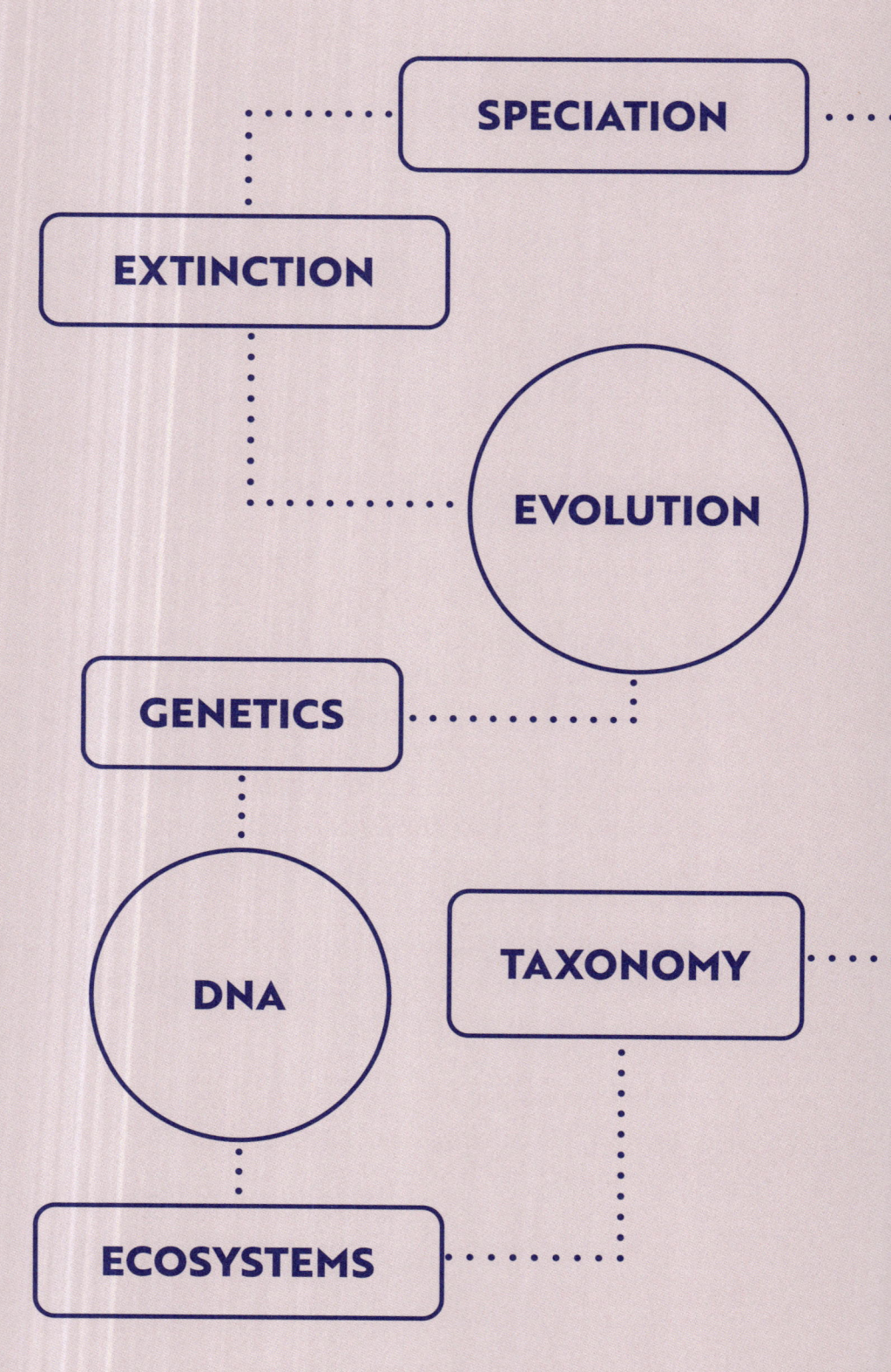

CHAPTER 1
ORIGINS OF DIVERSITY

- BIOSPHERE
- GEOLOGICAL PROCESSES
- CONSERVATION
- HOTSPOTS

INTRODUCTION

Around 4.6 billion years ago, gas and dust swirling around the young Sun pulled together through gravity to form a fiery mass of molten rock. Early Earth was an unforgiving place. Asteroids pummelled our new planet as it churned itself through continuous **VOLCANISM**. Eventually it cooled, oceans formed and great tides of acidic water swept over the land. Gradually, Earth became more hospitable to life.

We may never know exactly how life first appeared: gradually coming together in a mixture of chemicals and water known as 'primordial soup', or in some kind of biotic big bang. Even the form is unknown, though one popular theory suggests life began as simple self-replicating strands of **RNA**. By 3.7 billion years ago and possibly earlier, **UNICELLULAR ORGANISMS** known as prokaryotes appeared. Some 2.5 billion years ago, cyanobacteria learned to harness energy from the Sun and released oxygen in the process. Around 2 billion years ago, one prokaryote found its way into another, creating the first eukaryote and paving the way for the evolution of **MULTICELLULAR ORGANISMS** like plants, fungi and animals – including ourselves.

That life exists at all on our planet is something of a miracle – at least life as we know it. The Earth orbits the Sun within a hospitable zone just right for liquid water. The star's heat provides the energy needed to power life. Our atmosphere is a finely held balance of nitrogen, oxygen, carbon dioxide and the other gases we need to breathe. Life in its many forms has survived monumental planetary changes, including great reformations of continental land; at least two periods known as **SNOWBALL EARTH** which saw the planet totally covered in ice; and yet more asteroids.

What role has life itself had in this remarkable success story? In the 1970s, the English scientist James Lovelock (1919–2022) and American microbiologist Lynn Margulis (1938–2011) developed the **GAIA HYPOTHESIS**. This proposed that the Earth is a self-regulating system – acting like a living organism – that maintains a dynamic yet stable physical and chemical environment suitable for life, from the salinity of the oceans, to the composition of the atmosphere, to the temperature on the surface. Working as one giant ecosystem, living and non-living parts of the planet combine into a thermostat, one which has calmed other planetary processes that may have erased life entirely.

The Gaia hypothesis was met with initial scepticism. Though it remains controversial, evidence has grown in support of the idea and it is now being investigated through the field of Earth system science. Gaian perspectives have gained fresh prominence in the face of anthropogenic climate change, which Lovelock suggested should be tackled, along with the destruction of biodiversity, as one problem – with a unifying solution.

ORIGINS OF DIVERSITY MAP

EVOLUTION

SPECIATION
Formation of a new and distinct species through the process of evolution which develops its own unique characteristics and is separated by some form of barrier.

THERMOREGULATION
Mechanism by which mammals self-regulate their own body temperature irrespective of the external temperature.

CHARLES DARWIN (1809–1882)
English naturalist, biologist and geologist responsible for the theory of evolution by natural selection, the foundation for modern evolutionary theory. Author of *On the Origin of Species* (1859).

ENDEMIC
Plant or animal that only exists in a specific area.

SPECIES RICHNESS
Number of species within an ecosystem.

CAROLUS LINNAEUS (1707–1778)
Swedish botanist and explorer who developed the system of taxonomy still used today.

SPECIES EVENNESS
Measure of the relative abundance of species in an ecosystem. It is highest when every species in a given sample has the same abundance.

TAXONOMY
Naming system used to identify and group living organisms, with eight levels (domain, kingdom, phylum, class, order, family, genus and species).

GENETICS

DNA
(deoxyribonucleic acid) Molecule that carries the blueprint of genetic information an organism needs to develop and function.

RNA
(ribonucleic acid) Molecule that uses the information stored in DNA to perform several crucial roles within a cell, including creating proteins.

CHROMOSOME
Long, thread-like molecule of DNA that carries genes, found in the nucleus of plants and animals, and in bacteria.

MULTICELLULAR ORGANISMS
Those organisms composed of more than one cell, such as plants, fungi and animals.

UNICELLULAR ORGANISMS
Those organisms composed of only one cell, such as prokaryotes and some eukaryotes.

BIOSPHERE
Part of Earth inhabited by living organisms and the global sum of all ecosystems, from rainforests and mountaintops to soil and deep ocean trenches.

GAIA HYPOTHESIS
Theory suggesting that living organisms on Earth interact with their non-living surroundings to create a complex self-regulating system.

BIODIVERSITY
The diversity of life on Earth, including genetic variability, species diversity and ecosystem diversity.

EARTH

BIODIVERSITY HOTSPOT
Describes a region or locality with a high biodiversity that is under threat, particularly one with many endemic species.

SNOWBALL EARTH
Hypothesis that Earth was covered by ice for two periods, about 640 and 710 million years ago, with each glaciation lasting around 10 million years.

GENE
Short section of DNA on a chromosome containing the instructions that determine a living organism's physical appearance and other heritable traits.

VOLCANISM
The eruption of molten rock from inside the surface of Earth.

How did the platypus get its bill?

→ **Speciation is the key to biodiversity. Nature works in intriguing and mysterious ways, creating new species that branch away from each other – and which end up in many different, fascinating forms.**

All life that exists today is the result of over 3 billion years of evolution. Once, the only inhabitants of our planet were microbes. The first animals appeared perhaps only 890 million years ago. Over generations, nature has crafted millions of species that split from common ancestors. Biodiversity is the variety and variability of these life forms.

Several key intervals mark the history of this diversification, including the Cambrian explosion (over 500 million years ago), a radiation of life that created most of today's major animal groups and forms.

Over 99 per cent of the estimated 4 billion species that have ever lived no longer exist. Biodiversity is an evolving balance between two processes: speciation and extinction. The world has seen five mass extinctions, with proposed causes ranging from plummeting ocean oxygen levels, to dramatic changes in climate, to the asteroid impact thought to have wiped out most dinosaurs. Scientists believe we are now in the midst of the sixth – this time caused by humans.

Speciation happens as organisms diverge to the point they can no longer create fertile offspring. This can happen due to physical barriers, say, or if a species spreads out over large areas. In time, speciation leads to new forms and behaviours, and organisms better suited to their environment. To date, scientists have recorded around 1.7 million species of animals, plants and fungi, and there could be up to 100 million yet to be discovered – possibly a lot more if we include microbes, which we should.

Species spread out like the descending branches of an evolutionary tree. The platypus is certainly unique and the result of many splits from previous lineages. Like a bird, it has a bill – filled with electroreceptors that let it find prey in dark water – webbed feet and lays eggs. Like a reptile, it produces venom. Yet with fur, warm blood and the ability to produce milk, it is undoubtedly a mammal and just one example of our planet's extraordinary diversity.

SPECIATION

Over time – sometimes in just a few generations, sometimes over millions of years – organisms split into separate groups, evolving new traits that make them unique, while maintaining some traits from their shared ancestors. Around 166 million years ago, a split in the evolutionary line sent egg-laying mammals one way and those with a pouch or placenta another. Then, 130 million years ago, the first monotremes emerged, and in the seasonal darkness of ancient Australia some developed a bill that could let them hunt in the dark.

How did the herring teach the smelt not to freeze?

→ **They didn't mate with each other, that's for sure. It seems there's a different, rather convenient transfer of genes that can happen across evolutionary lines, as is the case for smelt, which inherited an antifreeze gene from herring.**

⇗ Our understanding of speciation stems from the theory of evolution devised by Charles Darwin (1809–1882). In 1859, the British naturalist published *On the Origin of Species*, in which he proposed that evolution happens through natural selection. This theory was conceived independently at the same time by the English naturalist Alfred Russel Wallace (1823–1913).

Darwin's theory suggests that individuals within a species gradually become different to each other in certain ways. Some of these variations – in colour, size or the ability to fight disease, for example – make them more likely to survive and reproduce in their current environment, and pass on the traits to the next generation. Eventually, individuals with these traits will outlive those without them – survival of the fittest.

We now know these changes result from variations in genes. Sections of DNA molecules hold the genetic information an organism needs to develop and reproduce. Genes contain blueprints for creating proteins and determine heritable traits.

There are several main sources of genetic variation. For example, the process by which cells divide and copy DNA can be imperfect and introduce mutations, a major driving force for variation. Another source is sexual reproduction, which creates new genetic combinations. Gene variation can also arise through interbreeding between different populations.

In 1928, scientists discovered horizontal gene transfer, a means of imparting genetic information without reproduction. DNA can be taken up from the environment or passed through viruses. Horizontal gene transfer is most common among bacteria, but can happen between distant species, and even between animals and plants.

A classic example of horizontal gene transfer is that of the herring and the smelt. Both live in icy waters and have a gene to create antifreeze proteins to stop their blood freezing. It's thought the smelt borrowed this ability from the herring around 20 million years ago. This sort of genetic transfer could be far more common than previously thought.

DNA: THE BUILDING BLOCKS OF LIFE

DNA is a complex molecule that includes all the instructions we need to form life. It contains the genetic code and is present in almost every cell in our bodies. It uses a language based on four chemical bases, labelled A, T, C and G. The order of pairings between these bases creates an instruction manual for the cells. Genes are sections of DNA that serve specific functions. DNA is bundled up into chromosomes, which carry genetic information between cells.

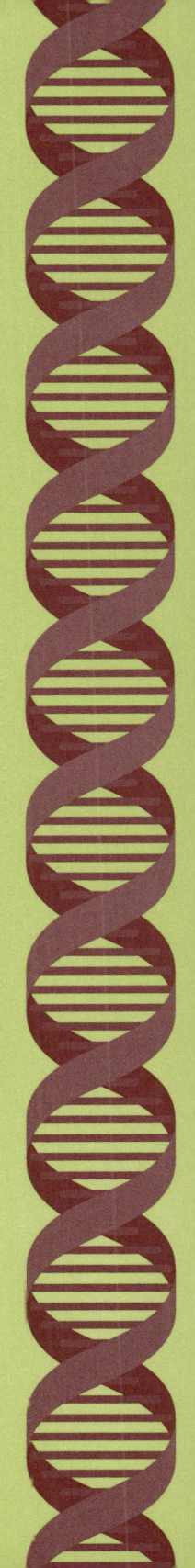

How do we tell a giraffe from a tree?

➡ **It might seem obvious to the naked eye – one is a large animal and the other a tall plant – but with so many species on the planet we've had to rely on scientists to help us tell them apart.**

↳ Scientists need a classification system to identify and explore all the different species on Earth. The first major effort came from Aristotle (384–322 BCE), a Greek philosopher who classified animals based on physical features such as whether they had blood or whether they laid eggs. There were several subsequent efforts to redefine the classification system, most coming after the 16th century.

The current system of taxonomy was developed by the Swedish botanist Carolus Linnaeus (1707–1778). It is still used today, albeit with extensive refinements. His naming system, published in the 18th century, categorised all animals and plants into kingdom, class, order, genus and species. This hierarchical categorisation also introduced binomial naming: scientists name organisms by their genus and species. The Linnaean System allows scientists to identify organisms and their evolutionary relationships.

The term 'biodiversity' – a contraction of biological diversity – was popularised by the American entomologist E.O. Wilson in 1985. Biodiversity is measured according to three scaled levels: genetics, species and ecosystems.

Genetic diversity is the variety of genes in populations and how they are expressed. Species diversity refers to the variation between organisms. Ecosystem diversity evaluates the composition of life in and between ecosystems such as tundra, forests and oceans – and the environments themselves.

The most common way to assess biodiversity is through species richness, the count of different species within a different area. Species diversity is a combination of species richness with species evenness – a measure of their composition. Areas with higher levels of species richness are considered to have higher biodiversity. Others use taxonomic diversity as a measure, calculating the distance or proximity of organisms based on their position on the evolutionary tree.

Advances in genomic sequencing technologies now let scientists detect, identify and count species with greater ease and accuracy. While potentially making future classification more complex, these technologies could improve our ability to both monitor and conserve diversity in our ecosystems.

TAXONOMY: THE ORDER OF THINGS

In Aristotle's hierarchy, humans were at the top. Now we know this isn't quite right. The current system of taxonomy has eight levels, from domain down to species. Organisms are put into groups based on their similarity to one another in form. The binomial naming system gives the last two ranks, genus and species, in that order. A wolf is *Canis lupus*, a dog is *Canis familiaris*. They're both canines, but you'd only want to invite one of them into your home.

Is there strength in numbers?

→ **Yes, and complexity is key. A rainforest teeming with trees, animals, insects and fungi all interacting is stronger than a monoculture of bamboo trees.**

An ecosystem is an interconnected community of organisms and their environment. Ecosystems are not static and the composition of species, and the interactions between them, change over time. Every species plays a role, however small, and the myriad interactions between them feed into the overall natural network. A simple example: microbes help plants access nutrients and plants transform sunlight into energy for insects, which are then eaten by birds that help spread the seeds of plants. The remains of the seeds are eaten by the microbes in the soil.

Just as the diversity of species between ecosystems varies, so too does the complexity of their interactions. Like any complex system, the higher the number of connections, the greater an ecosystem's overall stability. Imagine a patch of rainforest, teeming with millions of species of trees, animals, insects and fungi and all interacting with each other and the environment. Then imagine a monoculture of bamboo trees. Which do you think would be stronger?

The characteristics of an organism that serve a role in the ecosystem are known as functional traits. The more genetic diversity, adaptations and traits available in an ecosystem, the more likely it is to withstand external challenges such as disease, extreme weather events or climate change. The more coping skills available to deal with threats, the likelier an overall population is to survive.

An ecosystem is considered more resistant if it suffers minimal variation under shifting environmental conditions, and its resilience is based on how quickly and successfully it can return to its former state.

Not all environmental disturbances are necessarily harmful overall, however: a small adverse weather event can remove some species, freeing up resources for new ones to move in. This could end up increasing the overall complexity within an ecosystem. Extreme events like hurricanes or floods tend to be bad news for many organisms, though, and have the potential to unpick many of the interwoven connections within an ecosystem, thus making it weaker.

THE ECOSYSTEM WEB

An ecosystem is a complex web of interactions. Some are more helpful or harmful to certain organisms, but they all help to balance the overall whole. The existence of so many overlapping links is important because it provides different options. Owls, for example, can eat seed-eating birds, mice, squirrels or rabbits. If one of these species disappears, the owl will still survive. These relationships dive right down into the microbial level, with every organism playing its part.

How much is a coral reef worth?

⟶ **Putting a value on this is tricky. Economically, coral reefs and other ecosystems are worth trillions of dollars each year, but the natural world should be appreciated for its intrinsic value too – and in this sense it is priceless.**

Biodiversity is fundamental to the systems that support human life. Earth's ecosystems help to regulate our climate and provide us with clean air, clean water, food, fuel and medicines.

The extrinsic value of biodiversity takes into account the financial benefits from all the goods and services it provides. If money is the measure, then estimates suggest the services provided by ecosystems reach trillions of dollars annually – double the world's GDP. There is also huge potential for developing new medicines or technologies inspired by biodiversity – as long as we can protect this diversity and avoid the substantial losses created by overexploitation.

Considering biodiversity from a financial perspective can help generate investments and policies for conservation programmes, along with schemes to protect nature as a whole. Nevertheless, there is a lot of variability and uncertainty in biodiversity, not least because we have only recorded a fraction of the life that currently exists. Some organisations are supporting a move away from the conservation of individual species, or within protected areas, to landscape-scale conservation – safeguarding the ecological functions and services of large ecosystems over the long term.

Nature should also be considered according to its intrinsic value – its true worth is not the same as its price. A number cannot capture the full value of its contribution to our world; it has intrinsic value simply by existing. And it ties living organisms into our own value system, one based largely on economic growth, which is a factor in biodiversity loss. Natural spaces and biodiversity are aesthetically pleasing, good for our health, and have provided spiritual and cultural wealth – and habitats – for human societies for thousands of years.

Valuing biodiversity is a complex issue – and a hotly debated one at that. Yet in some ways, it can be simple. Watching a hermit crab move into a new shell, an orangutan feeding its baby or luminescent plankton glowing in the ocean at night – could you put a price on that?

THE CORAL ECONOMY

Coral reefs are some of the most important ecosystems on our planet. Linchpins of the marine environment, reefs host a quarter of all known species in the oceans. They support the livelihoods of around 500 million people around the world, in over a hundred countries. The goods and services provided by coral reefs – to fisheries and tourism, for example – are valued at over $11.9 trillion per year. They also protect coastal communities from hurricanes and flooding, which begs the question: How much are those human lives worth, too?

Do mountains make life more interesting?

⟶ **Yes, they do, but over the course of a very, very long period of time – think millions of years. Seismic and geological processes like volcanoes and earthquakes reshape our world and influence the climate, altering biodiversity.**

⇄ The biosphere is the sum of all life on Earth, a global ecosystem made up of many connected others. While holding a tremendous amount of life, the biosphere is just a thin film covering Earth's surface. Biodiversity is shaped by geological processes that are far larger and slower, and happening below the surface.

Our continents and oceans sit on a series of tectonic plates, rocky structures that float on molten rock below. Over huge timescales, these plates stretch, deform, diverge and collide with each other. This causes seismic activity, like earthquakes and volcanoes, and shapes the land and oceans. Mountains are born from a collision between two continental plates – this is why some ranges are still rising. These geological processes also impact the climate, another key factor shaping biodiversity.

The breaking apart of continents and the formation of mountains create physical barriers that can separate gene pools. Continental collisions create new land bridges over which organisms can migrate and mix. The distribution of continents and their physical features also affects the global climate, with knock-on effects for biodiversity. The rocks, soils and sediments churned up by geological activity provide the minerals and nutrients necessary to sustain life. These are more varied in mountainous areas, which account for around 87 per cent of all life on land – and are especially rich with life in the tropics.

Geological processes are also key to the creation of endemic species, irreplaceable organisms unique to specific areas. On islands, organisms are isolated and develop along their own evolutionary path. Mountain building and the creation of archipelagos have driven up biodiversity across the planet, through the creation of many new ecological niches. Think of the Amazon river, which starts as a trickle of water in the Andes and flows through vibrant rainforest out into the ocean, supporting and transforming life along the way.

Over 200 million years ago, all our continents were part of one supercontinent known as Pangea. Their separation led to the creation of many new habitats and more speciation, and biodiversity is far higher today as a result.

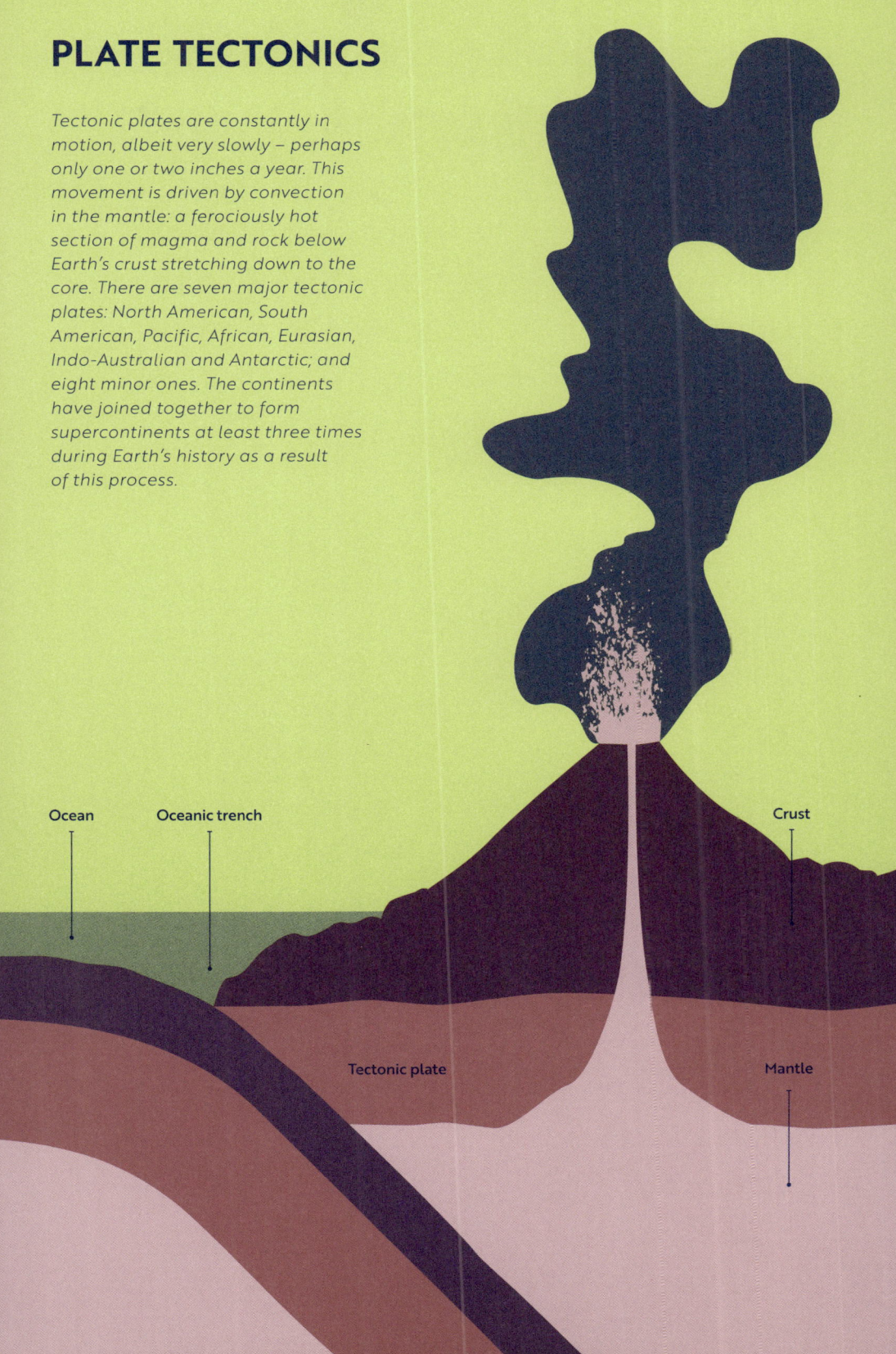

What makes a biodiversity hotspot?

⟶ Some parts of Earth seem to have it all, containing spectacularly high levels of biodiversity. These are known as hotspots, and they are – by definition – under substantial threat.

Biodiversity isn't constant around the world. Some ecosystems are considered 'hotspots', areas that contain an exceptional variety of life. To be designated a hotspot, they must contain more than 1,500 endemic plant species, and 70 per cent of the ecosystem's original vegetation must have already disappeared. The biodiversity hotspot concept was conceived by the British ecologist Norman Myers in the late 1980s as a way to draw attention not only to their beauty, but also their vulnerability.

There are 36 officially recognised biodiversity hotspots around the world, most of them in the tropics. In general, biodiversity increases towards the equator due to increasingly favourable conditions such as nutrients, heat and light. However, scientists still disagree on the reasons why biodiversity is richer in certain places.

Most biodiversity hotspots are in developing countries and are home to around 2 billion people, many of whom depend on them for their lives and livelihoods. Hotspots, as incredibly complex ecosystems, capture and store vast amounts of carbon from the atmosphere. This helps to mitigate the effects of climate change. Many scientists are concerned that due to human pressures and climate change the Amazon rainforest may be transforming from a net consumer of carbon into a producer, which could have devastating effects for the global climate.

The idea of biodiversity hotspots, while generally thought of as a positive step for conservation, has generated controversy. Some fear that dedicating outsized resources to these areas comes at the expense of less biodiverse parts of the world, which may be equally important to the healthy functioning of our planet. That being said, they are a useful way to check the health of the world's ecosystems in the face of limited time and resources.

We're also discovering that many regions thought to be barren of life, such as the deep sea or the Arctic, actually contain thriving ecosystems of which we were previously unaware. Life can be found pretty much everywhere – if you look closely enough.

THREATENED HOTSPOTS

Hotspots cover just 2.5 per cent of Earth's land surface but support almost 43 per cent of endemic birds, mammals and reptiles, as well as over half of the world's endemic plants. They are all impacted by human activities. While all are under threat, biodiversity hotspots are constantly evolving; some may recover, others may decline. Meanwhile, our knowledge, monitoring and conservation of biodiversity is improving. Scientists are detailing marine biodiversity hotspots around the world: these were left out of the original analysis but are equally important.

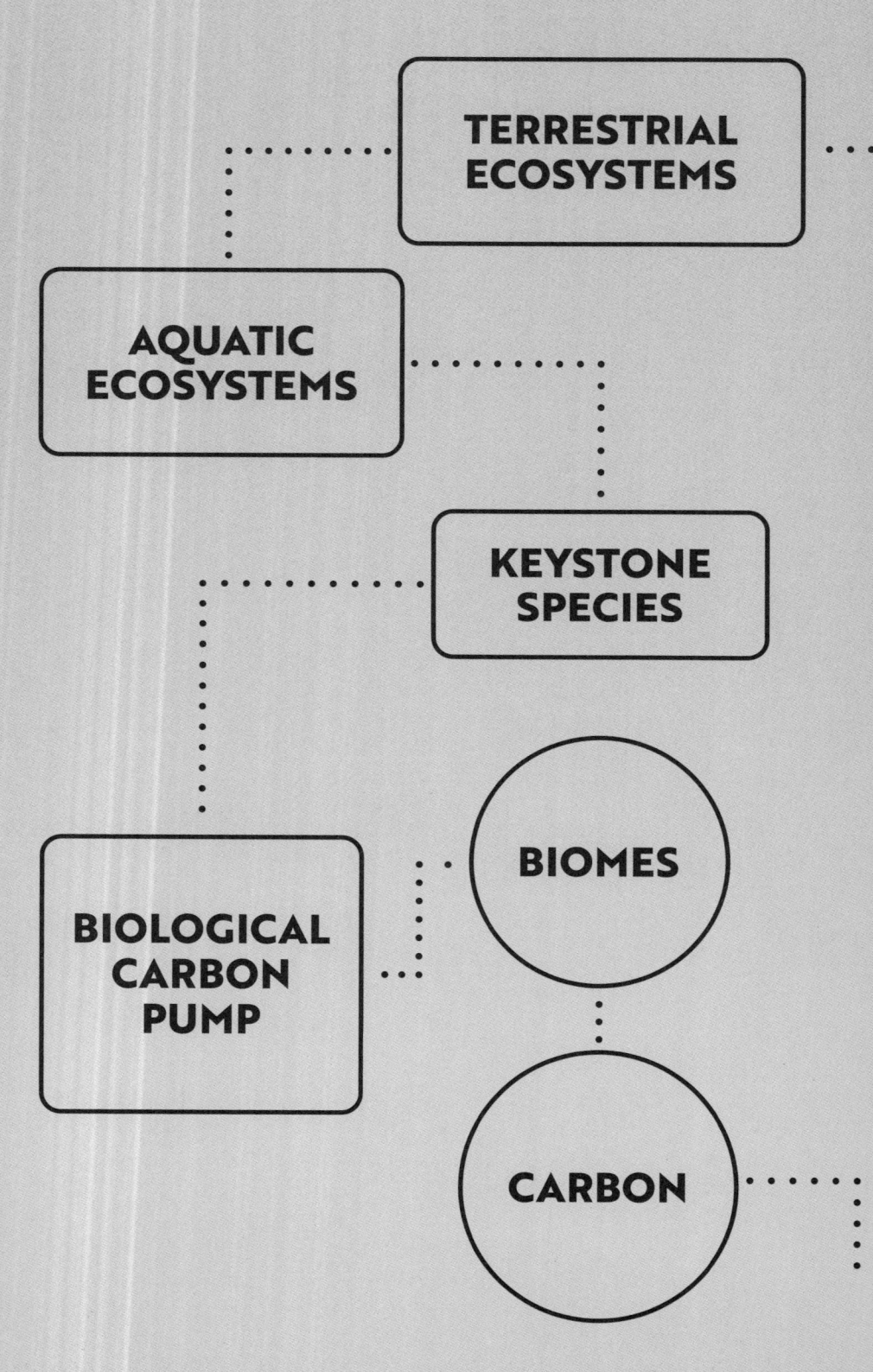

CHAPTER 2
ECOSYSTEMS

- ECOLOGICAL SUCCESSION
- SYMBIOSIS
- CONFLICT
- COMPETITION

INTRODUCTION

The word **ECOSYSTEM** stems from the Greek *oikos*, meaning 'household'. Ecosystems are self-sustaining ecological complexes of organisms and their environment, of living and non-living matter. These interplaying relationships can alter individuals, broader ecosystems and even the planet. The main energy currency exchanged in ecosystems is carbon. The interactions between species facilitate this exchange, which underpins biological life.

While ecologists have known about such interactions since the 19th century, the term ecosystem was coined by the British scientist Sir Arthur Tansley in 1935. Ecosystems, he wrote, are the 'basic units of nature on the face of the Earth'. He then explains that 'Though the organisms may claim our primary interest, when we are trying to think fundamentally we cannot separate them from their special environment, with which they form one physical system.' He continues: 'They form one category of the multitudinous physical systems of the universe, which range from the universe as a whole down to the atom.'

Then, in 1942, the American ecologist Raymond Lindeman (1915–1942) published research illustrating how energy and material flow through **TROPHIC DYNAMICS**. Using the example of a lake in Minnesota, he showed how **ABIOTIC** sunlight, through the chemical process of **PHOTOSYNTHESIS**, becomes **BIOTIC** plant tissue. Accumulated energy passes through herbivores and carnivores, then through a central collection of dead matter he called 'ooze', before being returned via bacteria and fungi to the biotic system. This already highlighted the importance of detritus within an ecosystem.

These ideas advanced through the work of Eugene Odum (1913–2002), who suggested that the world consisted of interlocking ecosystems and in the 1950s pioneered the field of ecosystem science. This explores the fundamental patterns, structure and functions within ecosystems and the forces acting upon them. It provides a scientific foundation for exploring questions ranging from the productivity of ecosystems to the factors underlying their degradation, as well as the loss of biodiversity and climate change.

Ecosystems thrive under a delicate balance of environmental conditions which can be affected by natural processes or human activity. When these conditions pass beyond a critical threshold known as a tipping point, the ecosystem can become irreversibly changed and undergo a shift or regime change into a new state, uprooting its former structure and internal dynamics.

Ecosystem science evolved during the second half of the 20th century, addressing international issues such as sustainable development and resource use. The Man and the Biosphere (MAB) Programme, launched in the 1970s, aims to unpick the complex relationship between humans and the biosphere, and has supported policies and innovations aiming to manage biodiversity. Ecosystem science fed into the United Nations' Sustainable Development Goals, a global initiative adopted in 2015 that aims to protect the planet and people on it. Ecosystem science is still evolving, and forming – with other fields such as atmospheric science and geophysics – an increasingly complex ecosystem of research.

ECOSYSTEMS MAP

CARBON

ABIOTIC
Describes the non-living physical and chemical parts of an environment (such as sunlight and soil) affecting living organisms and ecosystems.

CARBON CYCLE
The constant circulation of carbon between living and non-living things and the environment.

CARBON SINK
Place where carbon is stored away from the atmosphere, such as forests, permafrost and the deep ocean.

BIOLOGICAL PUMP
Set of processes by which carbon is drawn to the seafloor and sequestered away from the atmosphere, including through food chains and in falling carcasses.

PHOTOSYNTHESIS
Chemical process used by plants to make glucose and oxygen from carbon dioxide and water, harnessing light energy from the Sun.

ECOLOGICAL SUCCESSION
Process by which the composition of species within an ecosystem ebbs and flows, or changes completely – generally in the direction of more diversity.

NATURE

ECOSYSTEM
Biological system made up of a community of organisms in a particular environment, interacting with one another and their surroundings.

SYMBIOSIS
Association between two different organisms, whether mutualism (both species gain), commensalism (one species gains) or parasitism (one species harms the other).

SYSTEMS

BIOTIC
Describes the living organisms within an ecosystem, such as plants, animals, fungi and bacteria.

FOOD CHAIN
Sequence (usually diagrammatic) showing the feeding relationships between organisms. Every living thing is part of many food chains.

FOOD WEB
Relationships between different species, consisting of all the food chains in an ecosystem.

BIOME
Community of living organisms that has developed as a result of the physical environment and usually encompasses a large area.

TROPHIC CASCADE
Phenomenon in which an ecosystem becomes unbalanced due to the arrival or loss of an apex predator.

TROPHIC DYNAMICS
Theory pioneered by ecologist Raymond Lindeman describing the transfer of energy from one trophic level to another.

TROPHIC LEVEL
The position of a living organism in a food chain, food web or pyramid.

APEX PREDATOR
At the top of a food chain, the arrival or loss of which can have devastating consequences for an ecosystem.

COMPETITION
Rivalry between organisms over limited resources; between individuals of a species (interspecific), and between different species (intraspecific).

KEYSTONE SPECIES
Organism (usually a top predator) that exerts a major influence on an ecosystem, particularly one that is important to the survival of other species.

NICHE
Specific role played by an organism in an ecological community.

When does a river become an ocean?

→ It's a question of salt – obviously – and size. Rivers and oceans are both aquatic ecosystems, and everything is connected, with resources flowing from the land to the sea and vice versa, but the salinity divides them further.

An ecosystem is a system of interactions between biotic and abiotic, or living and non-living, things: animals, plants, fungi and bacteria all interconnect with rocks, air currents, rain, and so on. The boundaries between ecosystems are fuzzy – their influences spread far and wide. Ecosystems are everywhere; it's just a question of scale. A tree is an ecosystem, drawing life from its environment and providing a home for birds, bugs and bromeliads. When a zebra rolls on the desert floor, it forms an ecosystem in the sand. A river is an ecosystem. So is a small patch of earth by the river's bank.

Ecosystems are divided into two groups: terrestrial and aquatic. Terrestrial ecosystems are found only on land, whereas aquatic ecosystems are all about water, and include lakes, rivers, wetlands and oceans. These categories can be separated further. Freshwater ecosystems, like rivers, make up a fraction of aquatic ecosystems. Marine ecosystems, rich in dissolved salt, cover 70 per cent of the planet.

Ecosystems fit into larger patches of Earth called biomes. These vast mosaics are defined by animal and plant life and the regional climate. Land biomes are categorised according to their temperature and levels of light and rainfall, as these factors accommodate very different plants. Species diversity tends to increase in warmer, wetter climates – think tropical rainforests. Savannah biomes, which cover 60 per cent of Africa, have low rainfall, seasonal fires and grassy expanses grazed on by elephants and zebras. The subarctic taiga biome is home to thick, coniferous forests and well-insulated animals. The aquatic biome, the world's largest, contains all the oceans and water on land.

Back to our river, meandering through a forest to the sea. From its origin as rainfall in the mountains, it has carried oxygen and nutrients to countless ecosystems on its journey – aquatic and terrestrial. Salt flows into the ocean where it remains, and the freshwater ecosystem becomes marine. Waves crash to shore and feed nutrients into rock pools on land. Everything is connected.

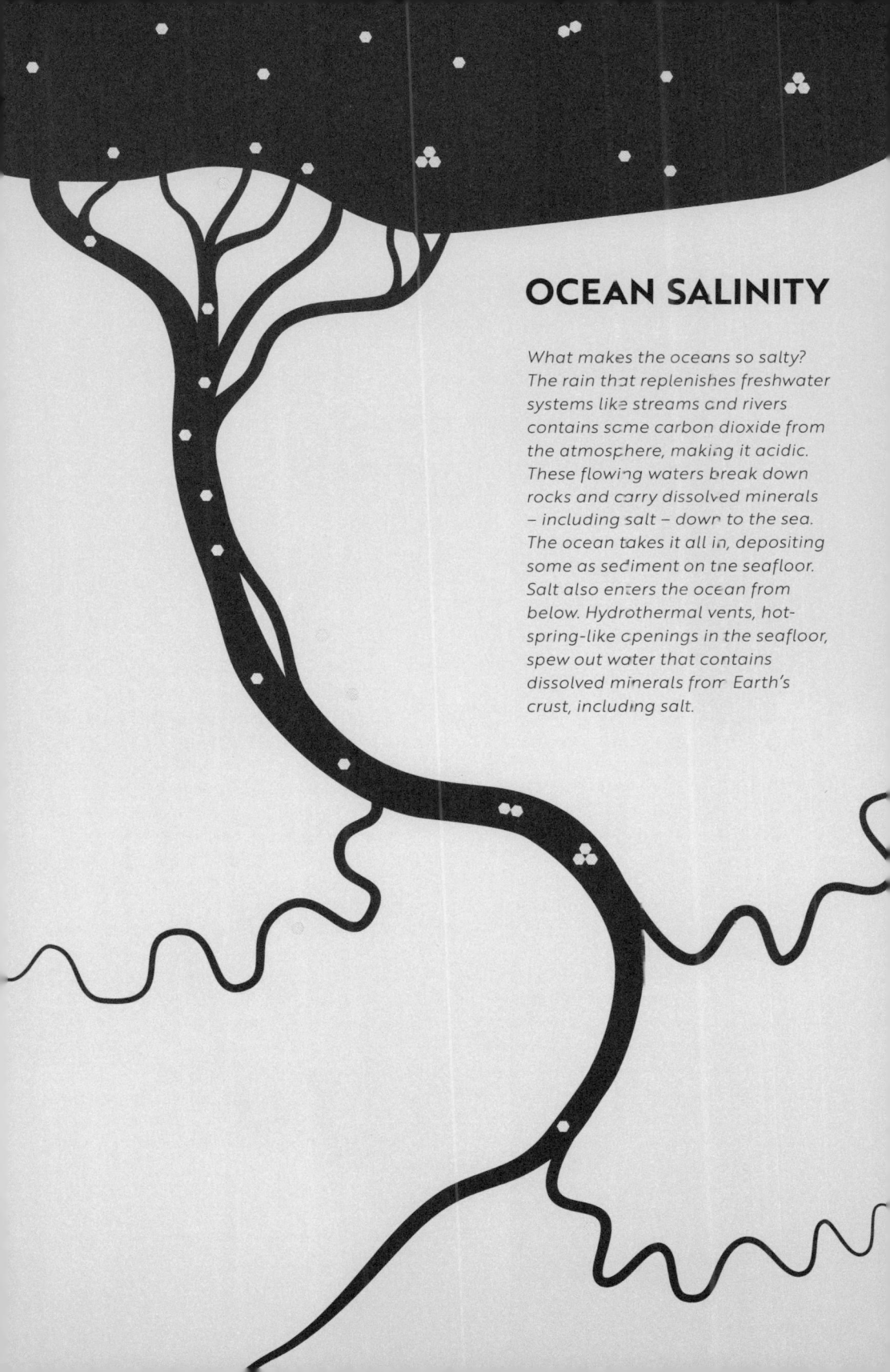

OCEAN SALINITY

What makes the oceans so salty? The rain that replenishes freshwater systems like streams and rivers contains some carbon dioxide from the atmosphere, making it acidic. These flowing waters break down rocks and carry dissolved minerals – including salt – down to the sea. The ocean takes it all in, depositing some as sediment on the seafloor. Salt also enters the ocean from below. Hydrothermal vents, hot-spring-like openings in the seafloor, spew out water that contains dissolved minerals from Earth's crust, including salt.

What happens when a whale dies?

➡ **In death, a whale can sustain many forms of life for over a hundred years. It feeds nutrients back into the ecosystem from which it came, which kick-starts new food webs from the ground up.**

⤵ Energy and matter are conserved within ecosystems and transferred between organisms. Matter is recycled, while energy flows through. When a whale dies, it falls through the water column, releasing energy and nutrients. First dibs go to sharks and seabirds if the whale bloats and floats on the surface. At the seafloor, scavengers like hagfish and hungry little crabs pick away at the flesh. Then bone-eating worms and microscopic organisms have their fill.

Nutrients and energy flow through ecosystems in a hierarchy of trophic levels. In most ecosystems, energy comes from sunlight. Through photosynthesis, plants or phytoplankton convert sunlight into sugars for other organisms – these energy creators are known as producers. Animals that get their energy from eating are called consumers. Herbivores – primary consumers – eat the plants, carnivores eat the herbivores, and omnivores like humans and bears eat both. Decomposers, such as bacteria and fungi, break down dead plants and animal carcasses, releasing nutrients back into the ecosystem.

Organisms form food webs across trophic levels, creating overlapping food chains of production and consumption. The more connections within and between food webs, the more resilient and productive the system.

While complex, this process is pretty inefficient: only about 10 per cent of the energy from one trophic level makes it up to the next. Energy is lost through heat, or as undigested food, or with animals that simply die of old age. Research suggests a greater biodiversity of both plants and animals makes ecosystems more efficient and productive – more energy in means more energy out.

The whale was once part of an ecosystem. As a concentrated source of food, its carcass essentially became one. Scientists rarely spot whale falls, yet love to study them. In recent years, several whale carcasses have been dropped intentionally. This is helping us better understand their effect on ocean health, climate change and even evolution.

WHALE FALL

Death creates extraordinarily complex ecosystems. When a whale carcass falls to the ocean floor at a depth greater than 1,000 metres (3,330 feet), it's referred to as a whale fall. On the seafloor, where nutrients are few and far between, whale bodies are a bonanza; animal carcasses tend to provide far more energy and nutrients than plants. In addition, research suggests they may also function as refuges and even evolutionary cradles for deep-sea organisms. Scientists can also use carcasses as indicators of an ecosystem's health.

What links a plant with a volcano?

→ It's carbon, the building block of all life on Earth. Carbon moves continuously around the world in a giant cycle, helped by both living and non-living things, including volcanoes and colliding tectonic plates.

Life on our planet only exists because of carbon. Carbon atoms are good at binding to others to form complex molecules, like proteins. When carbon bonds break apart, they provide energy for a range of biological processes. Carbon is constantly being transferred around the world between reservoirs of living and non-living things – from the atmosphere into oceans, soils, plants, animals and microbes – through the carbon cycle.

In the atmosphere, carbon normally takes the form of carbon dioxide or methane. The ocean absorbs carbon dioxide from the atmosphere, storing it in the deep as it cools and sinks. Places where carbon is stored away from the atmosphere, such as forests, permafrost and the deep ocean, are called carbon sinks.

Organic life cycles carbon through different reservoirs quickly. Plants take carbon from the atmosphere through photosynthesis, either storing it in the soil or returning it to the atmosphere as they decay. Animals take carbon from plants and release it into the atmosphere through respiration or when they die. Ecosystems are closely linked to the carbon cycle. The amount of atmospheric carbon will affect how much plant life can grow – but it also alters the climate system. As the oceans absorb more carbon they become acidic, impacting marine life.

The oceans, biosphere and atmosphere are all part of one giant carbon reservoir. Volcanoes and other tectonic processes – along with chemical reactions – cycle carbon over far greater time scales. Volcanic eruptions release carbon dioxide, as do tectonic plates when they collide. Most of Earth's carbon is stored in rocks and is released through weathering over time, eventually reaching the oceans. These fast and slow carbon cycles have stablised carbon levels throughout Earth's history.

Fossil fuels are remnants of carbon-based life buried underground for millions of years. Since humans started burning fossil fuels during the Industrial Revolution, we have dramatically increased the amount of carbon in the atmosphere. As this helps to regulate Earth's temperature, our planet is heating up – fast. Recent evidence suggests that animal biodiversity plays an important role in the carbon cycle, and more diversity (of large animals in particular) can help capture more atmospheric carbon. In other words, greater biodiversity makes for a more balanced planet.

BIOLOGICAL CARBON PUMP

Oceans hold 50 times more carbon than the atmosphere. Carbon is drawn to the seafloor as it cools in a process called the biological pump. Starting in phytoplankton at the surface, carbon then descends through ocean food chains. Carcasses head straight to the bottom. And each night, vast numbers of organisms migrate up from the deep to shallower depths to feast, then head back to the ocean twilight zone at dawn — known as diel vertical migration, this is the largest migration on the planet. Other organic matter falls through the ocean as marine snow.

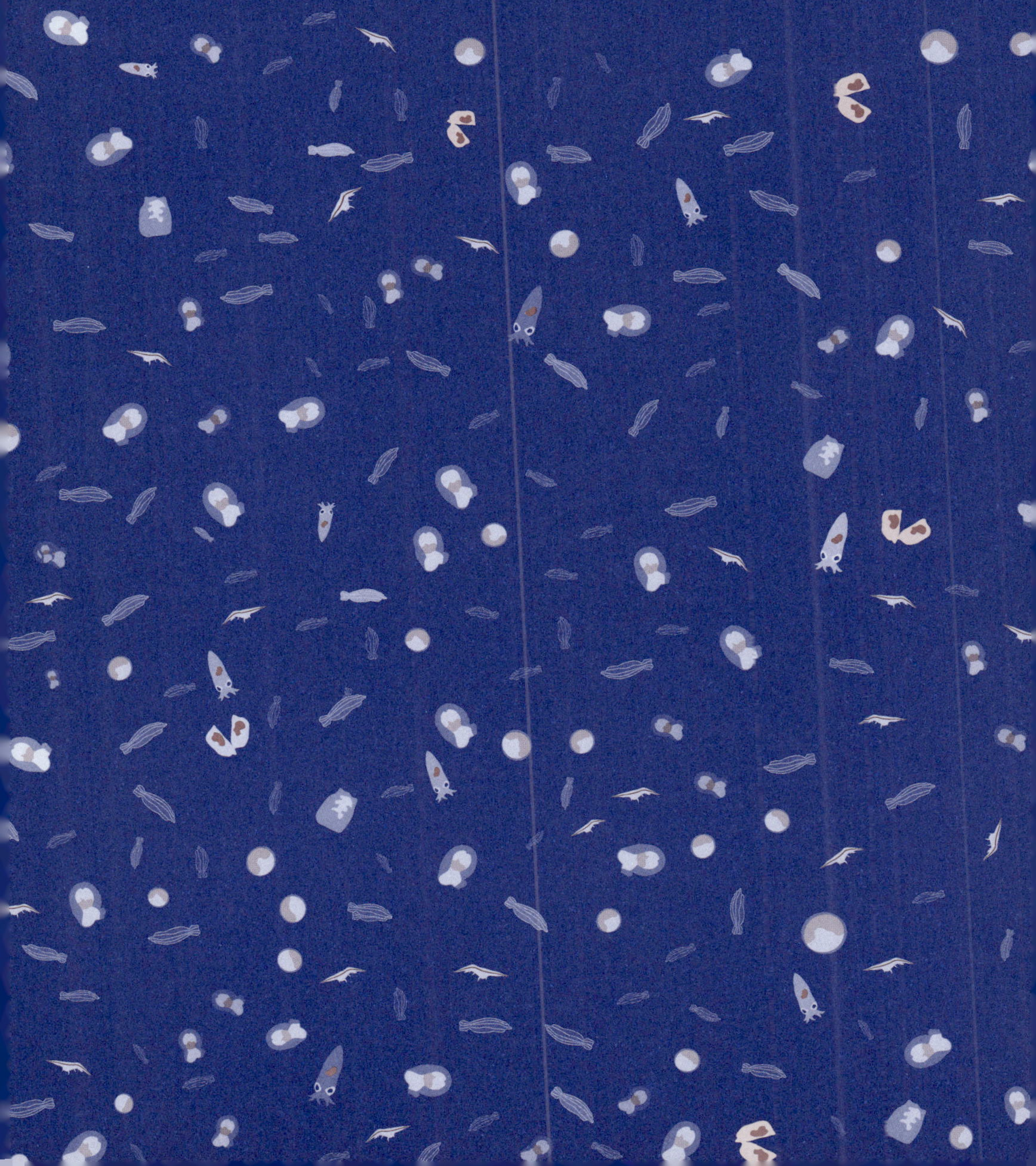

Is a shark more important than a scallop?

⟶ In a sense, yes. All species have an ecological role to play, but some define the ecosystem itself. Known as keystone species, they are often apex predators such as sharks.

While every organism in an ecosystem has a role to play, if a keystone species were to disappear, it would throw the entire ecosystem off balance – it would either change dramatically or collapse. Keystone species are often apex predators – those that enjoy a top spot in the food chain – but not always.

Predators feed on sick or injured prey, keeping overall populations healthy. They also bring balance to ecosystems through the indirect effects of their consumption spreading through food webs – a process known as a trophic cascade. For example, hammerhead sharks eat cownose stingrays, which feed on bivalves like oysters and scallops. In 2004, hammerhead populations collapsed off the eastern coast of North America. As a result, stingray populations exploded, decimating scallop populations and bringing down the North Carolina shellfish industry.

It's not all about predators, though. Prairie dogs are keystone species, too. These plant-munching little rodents burrow vast, elaborate underground colonies. Their tunnels provide shelter for rattlesnakes, jackrabbits and burrowing owls. As prairie dogs till the land, they churn up and fertilise soil, spawning vegetation to feed bison and elk. This makes them ecosystem engineers: another type of keystone species. Prairie dogs support over 130 other species, just by existing. Unfortunately for prairie dogs, they are also food for eagles, coyotes and a host of other animals.

Keystone mutualists are partnerships of two or more species that benefit each other. These are often pollinators: the plant provides food and the pollinator spreads the plant far and wide. The demise of one mutualist would impact the other, affecting the overall make-up of an ecosystem.

Back in the water, it's not all about the shark. Other fish can be keystone species, too. Think of salmon. Each year, salmon migrate upstream to spawn and die. They eat insects and regulate populations. Wolves, bears and birds eat the fish en route. Discarded salmon carcasses release nutrients, keeping soils fertile. Salmon make entire forests thrive, which seems pretty important.

TROPHIC CASCADE

An iconic example of a trophic cascade is in Yellowstone National Park, in North America, where, by the 1920s, wolves had been hunted to extinction. Elks thrived and destroyed plants, including willows, which beavers depend on for building materials during winter; and so beavers disappeared.

In 1995, ecologists started to reintroduce wolves, and not only are elk populations now more balanced, but as willows have grown in size and number, beavers have also returned. These ecosystem engineers are now building dams that help willow roots reach water and grow.

How does the butterfly deal with competition?

⟶ **After mating, some male butterflies use special chemicals to ward off other males to stop them competing to father offspring. Not all nature is fun and games: competition is rife; survival is paramount.**

⤻ With so many species and limited resources on our planet, there is inevitably stiff competition. Animals fight over food, water and space. Plants go to war over light, water, minerals and room for their roots. Competition is usually a lose-lose situation for those involved and only occurs when there are limited resources.

Every organism has a niche: a role within the ecological community. A niche provides the conditions an organism needs, including resources (like water) and interactions with other things (like prey). No two organisms can occupy the same niche for long – eventually they will compete and one species will drive the other to extinction.

Interspecific competition happens between individuals from different species. Fighting is more fierce, as the contenders occupy the same or overlapping niches. Drought in Zimbabwe, for instance, is increasing conflict between some elephant populations and humans. Intraspecific competition happens within species, such as two deer clashing antlers over a mate – the winner will pass on their genes to the next generation.

Conflict is everywhere in the natural world. Predators eat prey. Siblings attack one another, while parents and offspring may kill each other for their own benefit. Wasps usually follow their queen's orders, but scientists have found that some worker wasps in Costa Rica will revolt and kill their leader to shake up the colony's genetic make-up.

Sexual conflict happens when two sexes of the same species have different evolutionary goals: male butterflies want to mate as often and widely as possible, while females are more selective. After mating, male postman butterflies leave an anti-aphrodisiac pheromone on their partner to stop other males showing interest in her, which isn't great for the female.

Climate change could be driving up conflict between species, particularly in more extreme environments with scarce resources. For example, a war is breaking out between mountain goats and bighorn sheep over access to mineral licks – deposits of salt and other essential minerals – in the Rocky Mountains. So far, the goats appear to have the upper hoof.

COMPETITION FOR RESOURCES

Charles Darwin was the first to suggest that abiotic components such as climate are potentially more important to species conflict and survival in extreme environments, such as mountains. Glaciers around the world are melting due to climate change, opening up new resources like mineral licks. Highway building is restricting access to these resources, potentially driving up interspecies competition. In deserts, elephants are winning battles over scarce water. As access to drinking water continues to decline, will human conflict increase, too?

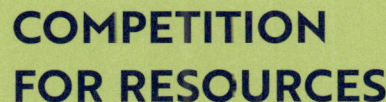

Would you want to hitch a ride on a shark?

→ **Perhaps not. But a remora would. These fish attach themselves to apex predators like sharks and manta rays, feeding on scraps as they travel the oceans together in a symbiotic relationship.**

↳ The natural world is not all dog-eat-dog competition. Many species live in symbiotic relationships, which benefit either one or both organisms.

Mutualism is when both species gain from co-living. A good example: microscopic algae called zooxanthellae that live within corals, illuminating them with vibrant colours. Corals shelter the algae, offering a protected environment and nutrients for photosynthesis. As a thank you, the algae create oxygen and remove coral waste.

Remora fish use a sucker-cup-like disk on top of their heads to latch on to sharks and manta rays. The hijacking remoras eat leftover food particles that float by as their host eats. This is called commensalism, where one species benefits and the other is neither helped nor harmed. It's unclear if the sharks get annoyed by the clingy remora, however.

The third type of symbiosis is parasitism, where one species harms another. In the succint description of E.O. Wilson, parasites are 'predators that eat prey in units of less than one'. Pity the poor crab, which can find itself host to the terrifying parasitic barnacle *Sacculina carcini*. Microscopic *Sacculina* larvae inject themselves into the crab's blood. The parasite grows, takes control of the crab's body and mind, destroys its genitalia and makes it host to a new parasite brood. The crab releases the larvae, even stirring the water with its claws to help them disperse.

You may not have noticed, but humans are in a range of symbiotic relationships. We have domesticated crops that sustain us. In doing so, we have spread these plants all over the planet. We raise honeybees that pollinate our flowers and fruit trees. We also shelter trillions of microbes on our skin and inside our bodies which help us survive.

Symbiotic relationships can be used to gauge the health of an ecosystem. These relationships have formed over millions of years, and can shape not only the physiology of organisms but also their interactions with the environment. Symbionts have a leading role in shaping biodiversity across the planet. Understanding them helps us protect biodiversity and conserve habitats.

PHORESY

The scientific term for organisms that hitchhike by latching on to others for a ride is phoresy. This phenomenon has been known since ancient times, and it's pretty common in the animal kingdom. Ticks, mites, fungi and remora fish all do it, and these phoronts – as they are known – have developed a range of methods to do so. Pseudoscorpions are phoronts. Pseudo or not, you wouldn't want one of those tagging along.

Why haven't sea turtles taken over the world?

→ **Baby turtles face a lot of threats: only one in a thousand makes it through to adulthood due to a range of factors such as predators, fishing and pollution that restrict their density.**

As we've learned already, most systems in nature are dynamic. Ecological communities of various species ebb and flow over time, and sometimes they change completely and replace those that came before. This process is known as ecological succession.

New life can form when an area of barren land is created, through volcanic eruptions or glacier retreats, or after an extreme disturbance. This is called primary succession. When lava cools and hardens, early pioneer species such as mosses and lichens arrive. Over time, plant diversity increases, moving through grasslands to larger shrubs and on to trees. Biodiversity grows more complex until it reaches a climax community, a complex ecosystem in equilibrium, such as a redwood forest. Secondary succession happens when land is disturbed but not enough to remove all previous life, as is the case after a forest fire.

Established populations also change over time. Growth would be exponential for species that have limitless resources and no predator to control their populations. But populations are limited by interspecies competition: eventually predators will arrive to control their prey. Predator and prey numbers then fluctuate in cycles.

Species are also limited by their ranges. This might take the form of physical barriers, such as oceans or mountains, or human infrastructure. Or a range may be limited because one species is in a mutualistic relationship and cannot spread without its compadre. Ranges are mostly defined by climatic conditions.

Sea turtles have existed for over 110 million years. But they haven't overrun the planet – or even the oceans, for that matter. Hungry crabs and birds prey on them, as do humans. Illegal trade in turtles, commercial fishing and plastic pollution pose new threats to their survival.

But it's not all bad news. Studying the rise and fall of ecological communities can help scientists conserve or restore damaged habitats. Management techniques like controlled burning in forests can also help keep ecosystems strong and biodiversity high.

DARWINIAN DEMONS

Life faces evolutionary restraints. These are known as trade-offs: one trait cannot advance much without worsening another. A common example is the rate of reproduction and lifespan. Animals that reproduce more tend to live shorter lives. If these barriers weren't in force, organisms could in theory improve to the point where they could live in any environment, give birth straight away, make many offspring and never die. These hypothetical beasts are known as 'Darwinian demons' – and they can't exist.

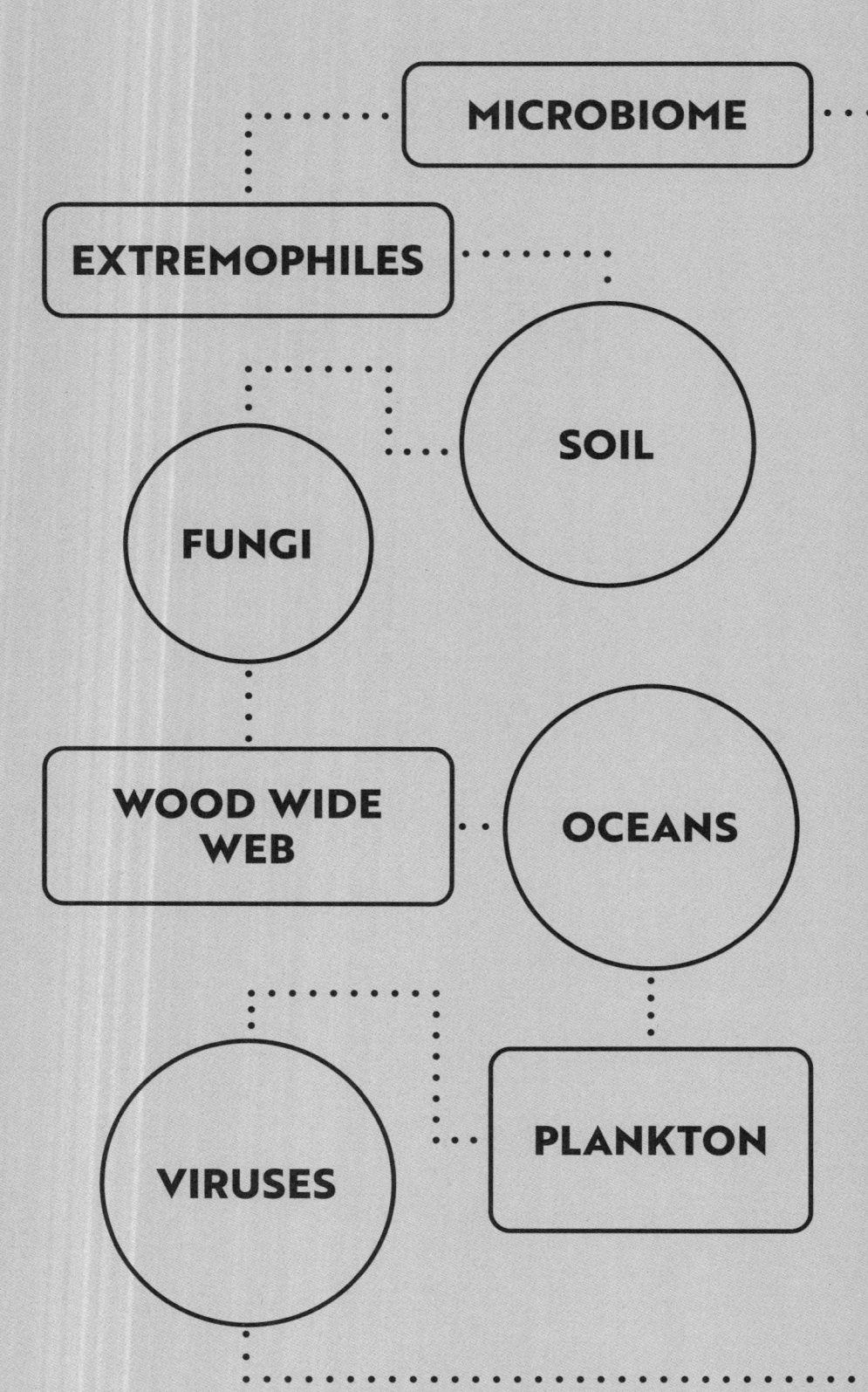

CHAPTER 3
THE UNSEEN WORLD

- MICROBES
- THE VIROSPHERE
- GENETIC TRANSFER

INTRODUCTION

Antoine van Leeuwenhoek's first love was fabric. A Dutch textile merchant, van Leeuwenhoek (1632–1723) crafted increasingly powerful lenses so he could assess the quality of his threads. Using these early microscopes, he discovered a new world filled with tiny organisms. Van Leeuwenhoek described many microbes, which he called animacules. He was the first person to see bacteria in plaque from his own teeth, writing in 1683: 'I then most always saw, with great wonder, that in the said matter there were many very little living animalcules, very prettily a-moving.'

Earth is a microbe world, and we're just living on it. Microorganisms were the first life on our planet and they will be here long after humans have gone. The Greek philosopher Anaxagoras (500–428 BCE) proposed in his **PANSPERMIA HYPOTHESIS** that 'seeds of life' can travel between worlds. In 1903, Swedish scientist Svante Arrhenius suggested that microbes could be pushed through space by radiation from stars. These ideas remain fringe, but are now being investigated more seriously.

Through chemical mastery, microbes have brought our dead planet to life. The Earth's **MICROBIOTA**, of bacteria, algae, fungi, **PROTOZOA**, **ARCHAEA** and viruses, regulate biogeochemical cycles on Earth, helping to transfer water, oxygen, carbon, nitrogen, sulphur and phosphorus. Astrobiologists searching for life on **EXOPLANETS** seek out chemical signatures that may suggest extraterrestrial microbes have performed similar tricks elsewhere.

Exactly how many microbes live on our planet is still debated. But they almost certainly make up the majority of biodiversity on Earth. There are an estimated 7.7 million species of animals, and up to one trillion species of prokaryotes (bacteria and archaea). The total count of microbial cells could be around a quintillion (10^{30}): a figure greater than the number of stars in the known universe.

Where are they all? A microbial tenet proposed by the Dutch botanist Baas Becking (1895–1963), 'everything is everywhere, but the environment selects', suggested that all microbes are cosmopolitan and could live anywhere. Now they are thought to exist in varied communities across certain parts of the world. Humans, households and cities all have a unique **MICROBIOME**.

Scientists suggest we are still profoundly ignorant about microbial diversity. The more we learn about microbes, the more we realise how indispensable they are. Emerging and concerning evidence suggests that they are dying around the world – fast. The fungi in Europe's **MYCORRHIZAE** may have declined by 45 per cent in a century, as witnessed by the disappearance of their fruiting bodies.

Researchers are working in earnest to map, sample, describe and conserve microbes. Microbial life is now influencing ecosystem restoration: soil transplants to restore underground microbiomes are showing promising early evidence. Microbes form a life-support system for our planet, which we are now working to bolster from the bottom up.

THE UNSEEN WORLD MAP

MICRO

MICROBIOME
Collective term for the microorganisms, as well as their genomes, which live symbiotically on or within larger hosts.

HOLOBIONT
Collection of different organisms composed of a host species and its associated symbiotic microbes, considered as one evolutionary unit.

INTERSTELLAR

EXOPLANETS
Planets found beyond the reaches of our Solar System.

PANSPERMIA HYPOTHESIS
Theory proposed by Greek philosopher Anaxagoras (ca. 500–480 BCE) that 'seeds of life' can travel through space.

MYCORRHIZA
Extensive underground network of symbiotic fungi in soils through which plants communicate with their environment and one another.

RHIZOSPHERE
Biodiverse part of the soil found close to the roots of plants that contain millions of microbes.

CELLULAR

MICROBIOTA
Huge collection of bacteria, viruses and other microbes.

VIROSPHERE
Term used to describe the world of viruses, but now also references the various viruses inhabiting host species and environments.

PROKARYOTE
Unicellular organism such as bacteria without a nucleus (the DNA is found in the cytoplasm).

ARCHAEA
Unicellular prokaryotic microorganisms similar in size and structure to bacteria.

EXTREMOPHILE
Organism able to survive in extreme environments once thought inhospitable to life (such as hydrothermal vents).

EUKARYOTE
Organism with cells containing a nucleus, such as plants, animals, fungi and many unicellular organisms.

PROTIST
Unicellular eukaryotic organism that is neither animal, plant nor fungus.

PROTOZOA
Animal-like protist that feeds on organic matter such as other microorganisms or organic debris. Examples include amoeba and euglena.

MICROALGAE
Photosynthetic organisms that live in aquatic environments.

MICROBIAL LOOP
Describes a trophic pathway in aquatic systems in which dissolved organic carbon moves from microbes to higher trophic levels.

PHYTOPLANKTON
Microscopic photosynthetic organisms comprising bacteria, algae and plants that form the basis of the aquatic food web.

Should I shower every day?

→ **Perhaps not. Humans are home to microbes that live everywhere on and inside us. In fact, half of our bodies are made up of microbes. The food we eat can alter this 'microbiome', and so too could our daily cleansing.**

⇥ An extraordinary number of microbial communities live on and within us and other animals and plants. The collective term for micro-organisms that live in symbiosis with larger hosts – more specifically their genes – is the microbiome. It's increasingly clear that our microbiome is critical to our physical and mental health.

We're actually more microbe than human. Research suggests over half of the cells that make up a human are microbial. Fungi, bacteria, archaea, protists and viruses live in a smorgasbord of colonies in our guts, our mouths, on our skin, and in other niches. We acquire these microbe communities from our environment, many of them during childbirth, and their composition changes throughout our lives.

Our microbiome is crucial for our health, affecting immunity, nutrition and ageing. A growing body of evidence suggests that an imbalance in our gut microbiome, known as dysbiosis, is linked to a wide range of physical and mental conditions, including obesity, cancers, autism and Alzheimer's.

An unhealthy skin microbiome may let in new allergens, leading to the development of allergies.

The food we eat can alter the composition of the microbes living within our gut. These new microbes are more likely to crave similar food. So eating burgers and hot dogs will likely attract communities that want more unhealthy stuff. Washing our skin with soaps and other cosmetic products may impact our skin microbiome, too, though the science is still a little … muddy.

Just as other ecosystems are affected by natural or man-made disasters, so too can our microbial communities suffer from external effects. Antibiotics act like a nuclear bomb for harmful microbes, but those that help us can become collateral damage. Overuse of antibiotics is also leading to resistance among bacteria.

A related emerging therapy is faecal transplants. These procedures involve the placement of stool from a healthy human into the gut of another. Faecal transplants have shown promise, but are not yet widely used. Not to be tried at home.

THE HOLOBIONT

The discovery that humans, other animals and plants all share symbiotic relationships with vast communities of microbes has led to a new term in biology: the holobiont. This word encapsulates the idea that organisms exist not as individuals, but in partnerships that have co-evolved over millions of years. The holobiont is a complex ecosystem that includes the host, its microbial colonisers, and the whole set of interactions between them. 'Who am I?' suddenly seems like a far more complicated question to answer.

What makes snow turn red?

→ **Microbes! Some microorganisms can inhabit seemingly inhospitable places. These 'extremophiles', as they are known, include colourful algae that live in snow, and change its colour.**

Some microorganisms exist in environments that might seem uninhabitable for life. These extremophiles either depend on or tolerate hostile conditions. They live all over the planet: from the world's tallest volcanoes, to the fiery chemical waters of deep-sea hydrothermal vents, to the hyper-salty lithium pools of the Atacama Desert. Psychrophiles — those that like the cold of snow or glaciers — are found on every continent, and include the microalgae that live on the slopes of Japan's Mount Asahi, turning them green and red.

Extremophiles are widespread primary producers, supporting entire food webs in extreme, nutrient-poor environments like deserts or snow, where life may not otherwise survive. Like all microorganisms, they also play fundamental roles in cycling chemical elements, including carbon, nitrogen, sulphur and oxygen, through the environment.

As they have evolved unique molecular adaptations to life in extreme environments, these hardy microbes have great potential in medicine and biotechnology. Enzymes from bacteria in thermal springs helped us develop faster PCR testing, which was widely used during the COVID-19 pandemic. Some extremophiles have been used to create commercial biofuels. Scientists think radiation-resistant extremophiles could be deployed in spacesuits to protect astronauts from UV rays; others could one day help with off-world agriculture.

Extremophiles are pressing the known boundaries for the limits of life. The microbe *Geogemma barossii*, for example, lives at a cool 121°C (250°F), while other species dwell in pH0 acids – not so pleasant for us. Their remarkable qualities are helping scientists research how life may have started on Earth and where it could exist on other planets. Some extremophiles live in the cold, extreme deserts of Antarctica, in similar conditions to those on Mars. Many celestial objects in our Solar System are extremely cold. Our cold-loving psychrophiles may offer clues to astrobiologists in their search for extraterrestrial life.

ARCHAEA

Most extremophiles are archaea. These hardy microbes love unpleasant environments, like hot springs, but they live pretty much everywhere. The American microbiologist Carl Woese (1928–2012) discovered archaea in 1977, redrawing the evolutionary tree into three domains of life (bacteria, archaea and eukarya). Archaea look like bacteria, but are actually more similar to eukaryotes like us. Initially, Woese's ideas were met with a hostility similar to the environments inhabited by his new microbes. But, like archaea, his theories survived.

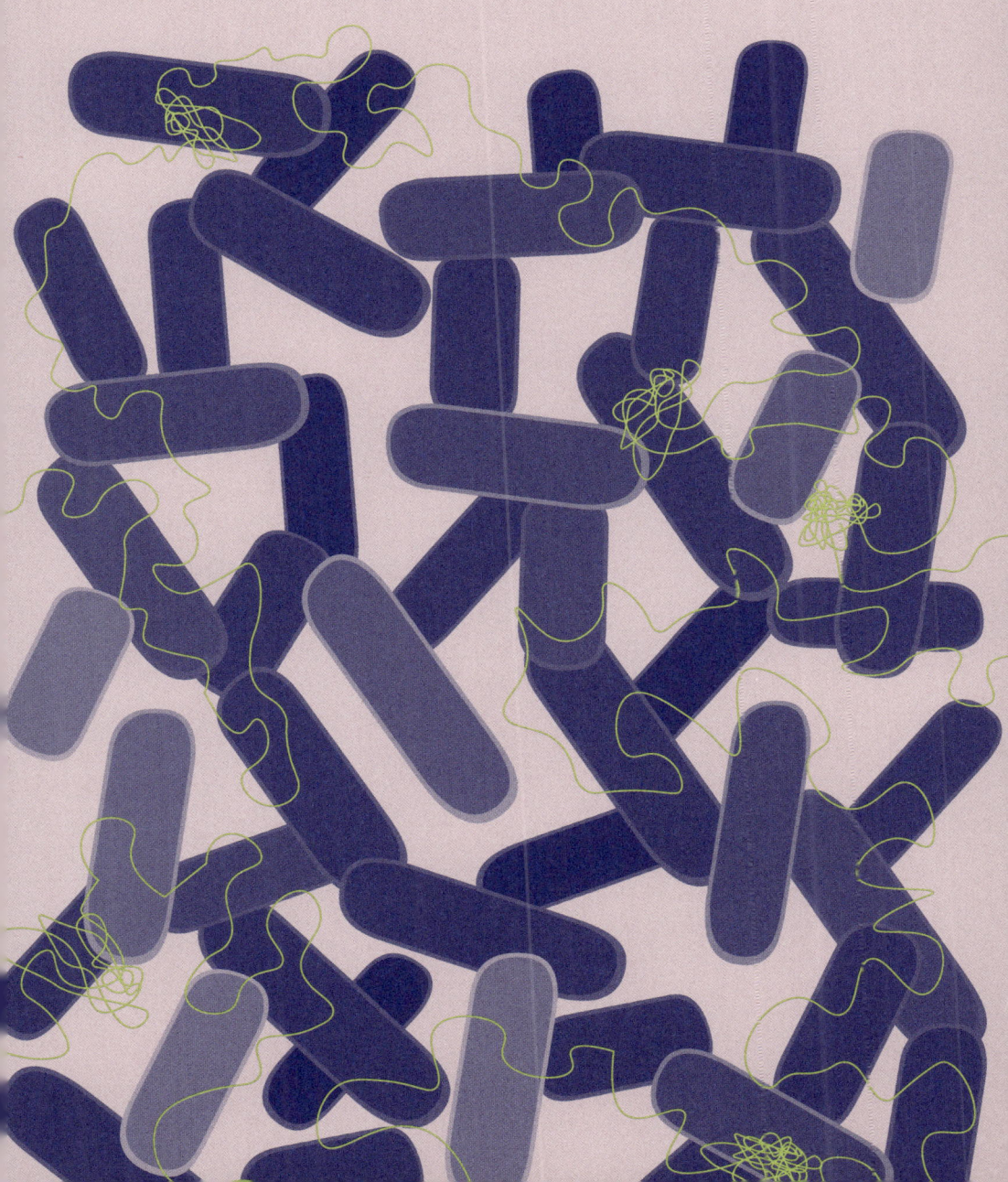

Could we survive another 'Dust Bowl'?

⟶ **We could. But as food systems are so interconnected today, a similar event would send ripples across the world, depleting global wheat stocks and harming the populations of many countries.**

⇶ Soil provides 95 per cent of our food. It has sustained human civilisation for millennia. Food systems and entire societies have crumbled when soil has not been maintained.

By the 1930s, decades of poor agricultural techniques had weakened soils across the Great Plains of the United States. As drought struck, strong dust storms swept away vast expanses of depleted topsoil, leaving behind millions of hectares of degraded, infertile land. The Dust Bowl, as it came to be known, caused cereal crop production to crash for a decade. Millions fled. A similar event today would have knock-on effects around the world, and deplete global wheat stocks by a third within just four years.

To the human eye, soil may seem like lifeless dirt. Look a little closer, and perhaps you'll see a worm, or a beetle, or even a burrowing mole surfacing briefly. Closer still, soil becomes an incredibly dynamic ecosystem filled with life, most of it microscopic. In just one gram of soil there are around a billion microbes, up to 50,000 different species of fungi, bacteria, archaea, protozoa and viruses all thriving below the surface. Soils hold over a quarter of the world's biodiversity. It's not just dirt.

This complex underground community – the soil microbiome – has profound effects on the planet. Soil microbes break down organic matter, recycle carbon, nitrogen and other nutrients, and support plant growth. They clean water in the soils by breaking down pollutants. Soil is the second largest carbon sink after the ocean, and the microbes' role in sequestering it is four times greater than any other factor. Soils hold more carbon than the atmosphere and all the world's plants combined.

A combination of intensive farming, pollutants, climate change and extreme weather is driving soil health to decline around the world. Our understanding of the soil microbiome and how it sustains this thriving ecosystem is only just beginning. Yet scientists are already exploring whether synthetic communities of microbes could offer solutions to degrading soils.

THE RHIZOSPHERE

One of the most critical and biodiverse parts of soil is the rhizosphere. This region closely hugs the roots of plants and contains up to 2,000 times more microbes than other soil. Plant roots secrete fluids containing sugars, amino acids and other compounds into the rhizosphere. This draws communities of microbes that exchange nutrients with the plants, which can help – or sometimes hinder – plant growth. Some scientists believe the rhizosphere may contain microbes that are beneficial to human health.

How do plants talk to each other?

→ **Fungi! Networks of fungi spread throughout the underground world, helping plants to exchange nutrients and 'chat'. Under forests, these networks are so large they're known as the Wood Wide Web.**

There is a lot of silent chatter beneath the ground. Plants communicate with their environment and even with each other through networks of fungi known as mycorrhizae, made up of long, thin chains called hyphae. Mycorrhizal fungi form a mutualistic symbiosis with plants, a relationship that has existed for over 450 million years.

Fungi receive sugars from the plants, who are treated to more nutrients and water from the soil in return. Plants can also pass chemical signals along mycorrhizal networks, communicating through the fungi with other plants and even with other species. If one plant is attacked by a herbivore, it can sound the alarm to others nearby. Defences are raised and everyone benefits.

Mycorrhizal networks are so extensive beneath forests that along with the tangled roots of trees and bacterial communities they are known collectively as the Wood Wide Web. Scientists have used AI to map the global extent of the Wood Wide Web. There are two main types of mycorrhizal networks. Some dominate tropical biomes and cycle carbon quickly, releasing it into the atmosphere. Others thrive in temperate regions and sequester more carbon into the soil – yet are more at risk from warming temperatures.

Fungi were long thought of as plants, but they are now recognised in their own incredibly diverse kingdom. They range from the single-celled to the complex organism, from microscopic spores and hyphae to the macroscopic mushrooms we see above ground (these are just the fruiting bodies of fungi living beneath). Fungi are pretty much everywhere in the living world. The yeasts we use to create bread and beer are fungi. They live inside our bodies. The world's largest known organism – *Armillaria ostoyae*, discovered in 1998 – lives in Oregon, and it's a fungus.

Only around 150,000 species of fungi have been described, yet estimates suggest there could be millions. Beyond applications in agriculture, scientists are exploring their use as natural biocontrols against vectors of disease like mosquitos, as environmentally friendly materials, and as indicators of ecosystem health.

FUNGI MIND CONTROL

Fungi affect the behaviour of plants and trees. These fascinating organisms can have powerful effects on other creatures too, such as controlling the minds of animals. Some fungi in the rainforests of Peru land on flies and infiltrate their blood. They take control of their host's mind, making it fly to a location ideally suited for fungal growth. The fly dies, the fungus feeds and grows, and spores are released to start the cycle once more. Gruesome and impressive.

How many microbes are in the ocean?

→ A lot. If you weighed all the mass in the ocean, over 90 per cent would be microbes. And they're important, playing a fundamental role in ocean food webs and other atmospheric processes.

Sailing on the open ocean, if you were to scoop out some water with your hands, you would be holding millions and millions of microbes. These floating microorganisms, along with larger plants and animals visible to the naked eye, are plankton. Their name comes from the Greek for 'wanderer': unable to swim horizontally, they drift along according to the whims of the ocean currents.

Phytoplankton are the forests of the sea. These photosynthetic organisms – bacteria, algae and plants – are critical to nutrient and atmospheric cycles. *Prochlorococcus*, the smallest and most numerous photosynthesising organism on Earth, produces roughly the same amount of carbon as global croplands. Diatoms, microscopic algae that live in aquatic ecosystems across the planet, including fresh water, breathe out more oxygen than all the rainforests combined.

The tiniest ocean microbes are the basis for all other life in the sea. Bacteria recycle nutrients from dead microbes and other organisms. Microbes called nanoflagellates graze on the bacteria and are in turn eaten by larger and larger organisms, through crustaceans and fish, right up to the largest whales. This process of microscopic cleanup is known as the microbial loop.

Microbes live everywhere, including the ocean floor, where they collect together in colourful mats. These microbes can tell us what chemicals are released from underneath the rocks below and which microbes live there too. With no sunlight, microbes here use chemical reactions to make food, from places like hydrothermal vents.

Ocean microbes also form symbiotic relationships, with each other and with larger organisms. Tiny, glowing bacteria live in the lure of the ferocious anglerfish, for example, drawing prey from the black ocean depths for their host to consume.

Some microbes have extremely short lifespans, so they are able to adapt quickly to changing conditions. Some bacteria are already evolving to eat plastic. Scientists hope ocean microbes can help us fight climate change by increasing the amount of carbon stored in the ocean.

NESTED ECOSYSTEMS

In the oceans, scientists are also studying holobionts in relation to their ecosystems. The transmission of microbes and nutrients through direct contact in the ocean can create widespread networks of interactions with knock-on effects. Sponges and their microorganisms, for example, can have substantial effects on the cycling of nutrients in their environment. This can alter the growth of other organisms, changing competition between species and ultimately the structure of communities in an ecosystem.

Are viruses a good thing?

⟶ **Well, they're certainly not all bad. Viruses have existed on Earth since the beginning of life, and possibly before. They have shaped the evolution of life on our planet, including that of our own species.**

Viruses are more like machines than living things. These tiny microbes infect the cells of organic hosts, deposit their own genetic code inside, and force the cell to replicate more viruses. Scientists still debate whether or not they are alive. They generally agree, though, that viruses are incredibly diverse, numerous and important.

There are thought to be more viruses on our planet than there are stars in the universe. They infect everything: people, animals, plants, fungi, microbes. They are in soil, in our guts, in volcanic springs, in the darkest reaches of the ocean. They have shaped biodiversity like almost no other force. As our most abundant organism, viruses may form the greatest reservoir of genetic diversity in the world.

The role of viruses is particularly evident in the oceans. Each day, marine viruses kill around 20 per cent of microbial life in the oceans. (There are ten times more viruses than bacteria in the ocean.) So viruses influence the composition of marine ecosystems, drive the biological carbon pump and profoundly affect global biogeochemical cycles.

Whenever a virus infects a host there is a possibility for genetic transfer. Viruses mutate, and when two viruses infect the same cell they can swap genetic notes. Viruses have affected the evolution of many species: over 8 per cent of the human genome is viral, genes that have affected the function of our brains and given us the placenta. Viruses pick up genetic information from their hosts too, shaping evolution in both directions.

Viruses are regarded as our nemeses: poliovirus, coronavirus, Ebola, to name but a few prolific human killers. Yet they can also be our allies. Scientists use viruses to fight cancer, and as a potential weapon against antimicrobial resistance, they could save many lives. The development of cheaper technology able to rapidly sequence genetic samples has opened our eyes to a world of viruses: the virosphere. The more we learn about this viral world, the more we realise how little we know.

THE VIROSPHERE

We are just beginning to explore the virus world: the virosphere. In the late 19th century, two scientists independently discovered the first virus, the tobacco mosaic virus. In 1938, scientists used an electron microscope to see the size and shape of a virus for the first time. Now we know that viruses have diverse methods for replicating and come in a range of shapes and sizes. They are composed of genetic material and are all shielded by a protein cloak.

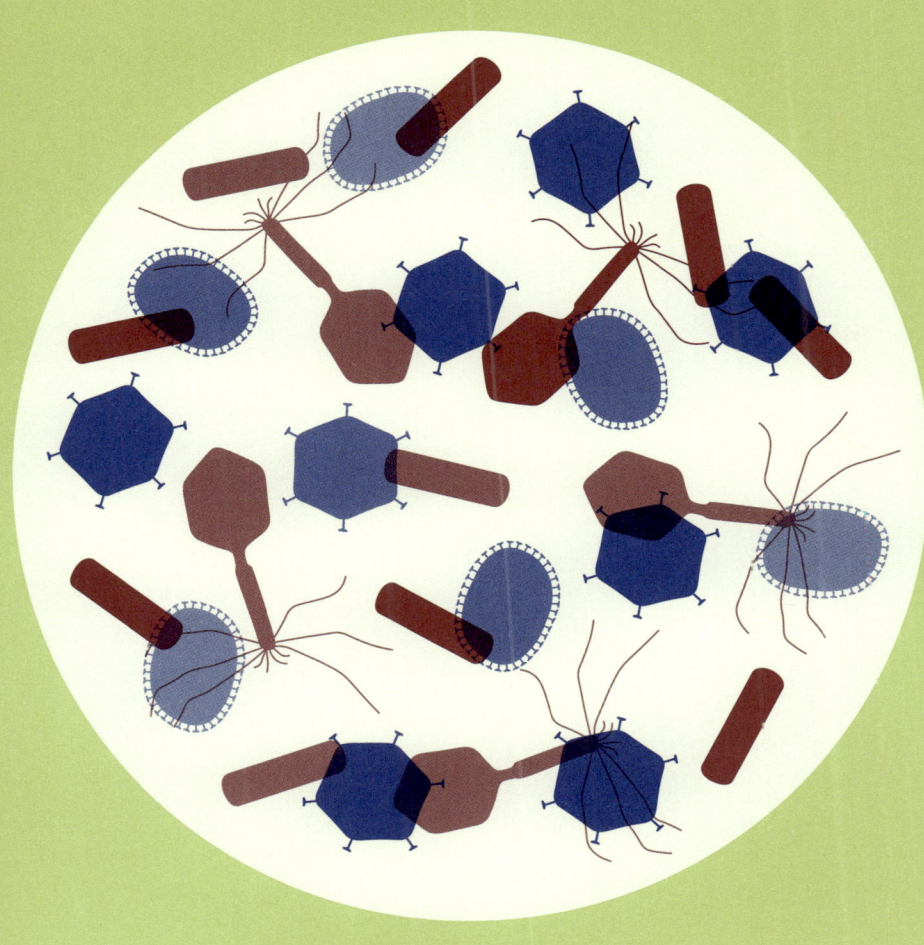

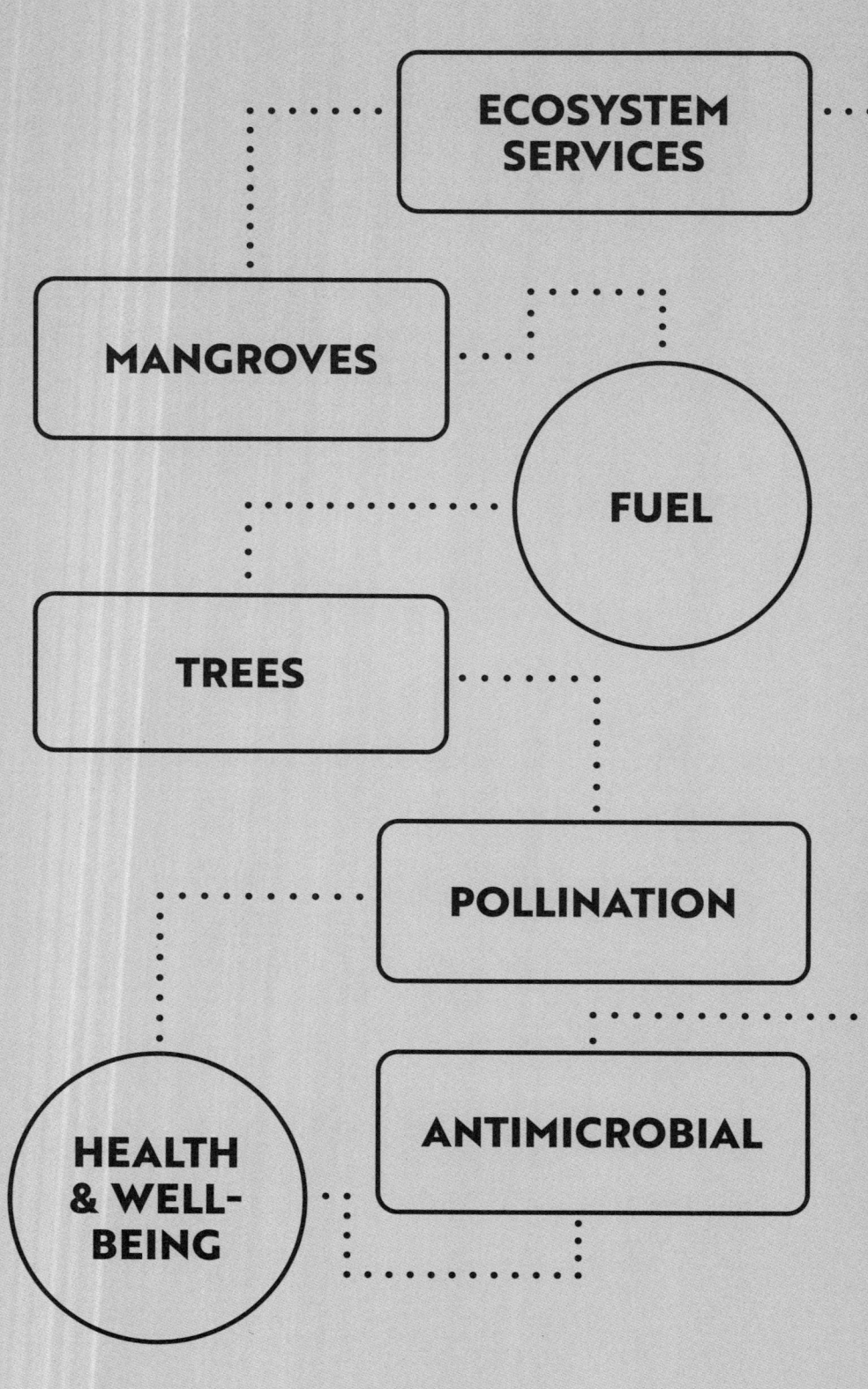

CHAPTER 4
OPPORTUNITIES

- PROTECTED AREAS
- SERVICES
- FOOD

INTRODUCTION

Nature's immense wealth has been spoken of since ancient times. So too has its disruption due to human actions. In 400 BCE, Plato described how deforestation was contributing to soil erosion and the drying of rivers and springs. This awareness likely extends far deeper into our past.

The goods or services, activity or function from ecosystems that benefit humans are known as **ECOSYSTEM SERVICES**. During the 21st century, scholars started to highlight the value of nature more explicitly and what could be lost. The ecologists Paul and Anne Ehrlich coined the term ecosystem services in 1981. Its definition has changed over time with the evolution of ecological and economic thought.

Initial assessments framed ecosystem services in a purely utilitarian manner. Some classical economists recognised their value for use, but considered them to be a free resource and available in quantities sufficient for all human consumption. As we now know, they are not endless or impervious to damage. Through the 1990s the notion of ecosystem services became increasingly mainstream. A landmark study in 1997 ascribed a value to global ecosystem services: up to $54 trillion per year.

Since then, studies exploring ecosystem services and how to value them have risen exponentially. As academic interest in them has grown, natural scientists have increasingly described ecological values in economic terms with the aim of generating public interest in the conservation of biodiversity. The concept has spread beyond academic circles and now influences international policy.

The list of benefits we gain from nature is extensive. Ecosystems regulate weather cycles, atmospheric gases, climate and disease. A forest can store carbon and provide food and timber, clean water and cultural identity. Around 4 billion people rely on nature directly for medicine. Ecosystems harbour genetic resources and generate healthy soils.

The perceived worth of these services can change over time along with our needs, even if the underlying ecosystem stays the same. Valuation is complicated because certain humans will place more worth on different ecosystem services. One community may favour harvesting a forest for timber, while another depends on the ecosystem's ability to provide drinking water or spiritual enlightenment.

Scholars recently proposed an alternative to **GROSS DOMESTIC PRODUCT (GDP)** to capture the value of ecological products and services to human society and well-being more accurately. Pioneered by the American biologist Gretchen Daily, **GROSS ECOSYSTEM PRODUCT (GEP)**, as it is called, has already been adopted in several pilots across China.

Nature and its services are deteriorating around the world. Some argue that putting a value on nature is inherently wrong, and that focusing on ecosystem services won't ensure that biodiversity itself is valued – particularly endangered species and their habitats. Others hope that valuing ecosystem services could foster a widespread cultural shift towards conservation of the natural world.

OPPORTUNITIES MAP

VALUE

ECOSYSTEM SERVICES
Goods or services, activities or functions derived from an ecosystem that are of benefit to humans.

ECOTOURISM
Mutually beneficial symbiotic relationship between humans and nature, with people gaining non-material benefits and their money helping to regenerate the natural world.

GROSS DOMESTIC PRODUCT (GDP)
Total monetary value of goods and services produced by a country in one year.

GROSS ECOSYSTEM PRODUCT (GEP)
The value of all ecosystem goods and services provided annually for human well-being in a given region.

ENERGY

BIOENERGY
Energy produced by organic materials – known as biomass – which contain the carbon absorbed by plants in photosynthesis.

RENEWABLE ENERGIES
Energies supplied by sustainable fuels and power, such as biofuels, solar power and hydropower.

BENEFITS

REGULATING SERVICE
Service provided by nature that is beneficial to human life, such as the sequestration of carbon by mangrove forests.

MANGROVE
Tropical forest that thrives in the salt water of coastal areas, rivers and estuaries, supporting diverse life both above and below the water.

MICROBE SYMBIONT
Photosynthetic primary producer living in symbiosis with plants, animals and fungi.

ANTIMICROBIAL RESISTANCE
The resistance of microorganisms such as bacteria and viruses against drugs such as antibiotics.

AGROFORESTRY
Emerging farming technique that integrates trees, shrubs, grasslands and animals with growing crops; a more sustainable approach that improves soil quality.

FOSSIL FUELS
Fuels from substances such as coal and oil produced as a result of earlier biodiversity: the compressed plant and animal matter accumulated over millions of years.

MONOCULTURE
System of agriculture in which one crop is grown in a field at a time, which reduces biodiversity and is ultimately less productive.

Why should I care about mangroves?

→ **Mangroves were long dismissed as just muddy ocean forests. But we now know that they provide a whole range of ecosystem services and goods that keep us – and other forms of life – alive.**

⇉ For an example of the riches nature offers, look no further than mangroves. These tropical forests thrive in the salt water of coastal areas, rivers and estuaries. They harbour biodiversity under water and above: from shrimp, to manatees, to fishing cats, to Bengal tigers, and many endangered and endemic species. This is a supporting service – one that helps to protect plants and animals, and foster genetic diversity. Supporting services are the basis for all other ecosystem services.

By housing marine creatures like fish and crabs, mangroves create food for humans. Tangible benefits that can be consumed, used or sold are known as provisional ecosystem services. By providing these, mangroves support the livelihoods of millions of people.

Mangroves are also custodians of the climate. They sequester carbon ten times faster than other tropical forests and store up to five times more in the ground. They clean water and air. These are regulating services: the processes of an ecosystem that maintain an environment that benefits human life.

This extends to physical protection. With roots that reach deep under water into the ground, mangroves form strong defensive structures. When hurricanes strike, the powerful roots and trunks slow winds and waters, and reduce flooding. They prevent over $80 billion of flood damage each year, and without doubt have saved many lives.

With their unique diversity of flora and fauna, mangroves also provide us with cultural services – non-material benefits, the fourth kind of ecosystem service – from tourism to spritual gains. Tourists are drawn to marvel at the biodiversity, or to fish or kayak nearby, while some communities in Fiji associate mangroves with deities.

A bridge between ocean and land, mangroves are one of the most important ecosystems we have. Yet they are under immense threat. Over a third have disappeared, and in some places they are vanishing faster than terrestrial forests. Long underappreciated, their role in fighting biodiversity decline and climate change is now clear. That should matter to all of us.

PROTECTIVE COASTAL ECOSYSTEMS

Many ecosystems offer essential shelter from natural forces such as hurricanes. Coral reefs soak up wave energy. Salt marshes reduce storm surges and prevent erosion. Sand dunes break up winds. Over 2.75 billion people – roughly a third of the world's population – live within 100 kilometres (62 miles) of the coast. In a future where extreme weather is predicted to rise, these natural barriers will only grow more critical to our survival. We could build our own infrastructure, but why not protect nature's own?

What fuels our survival?

→ A few things are essential for our survival: air, water, food and shelter. These days, we need a whole lot of fuel, too. Fortunately for us, these come from our ecosystems, which is why we need to protect them at all costs.

Plants take care of our first two fundamental needs: they clean pollutants and toxins from air and water, and exhale oxygen that we breathe.

Our food comes from plants and animals. Over time, we have improved the methods we use to take food from the natural world. But our basic needs remain the same. Healthy ecosystems create the optimal conditions for humans to grow, gather, hunt or harvest food. We use natural materials to build shelter, too. Humans constructed buildings with mammoth bones around 25,000 years ago, and our ancestors did so with wood around half a million years ago. Animal hides have protected us against harsh weather and climate for many thousands of years.

Yet there is perhaps no more transformative provisioning good in our history than fuel. The discovery – or control – of fire, perhaps over a million years ago, let us stay warm, defend against predators, light up the darkness and cook food, thereby avoiding harmful toxins in raw meat.

It allowed us to explore new terrains and alter our environment. The energy generated from burning organic material like plants is called bioenergy, and we've used a lot of it. It has shaped our evolution.

Ancient ecosystems still provide us with fuel. Coal and oil are products of prior biodiversity, plant and animal matter that has settled and compressed over millions of years. These fossil fuels are unevenly distributed across our land and oceans, shadows of former biodiversity hotspots. It is a sad twist of fate that the same ancient life that powered the Industrial Revolution is now the engine behind the decline in biodiversity today.

Now we are turning to more sustainable fuel. Renewable energies benefit from healthy ecosystems, which can boost water flow for hydropower, for example. We can even generate energy from microbes. Solar power was inspired by the photosynthesis of plants. There could be unknown fuel sources lurking throughout the natural world, and many others nature may inspire us to create.

SEMI-ARTIFICIAL PHOTOSYNTHESIS

Italian chemist Giacomo Ciamician (1857–1922), the 'father of photochemistry', first proposed the idea of recreating plant photosynthesis in artificial systems over a hundred years ago. With solar technology now well developed, scientists are adding natural elements back in.

The emerging field of semi-artificial photosynthesis incorporates photosynthetic enzymes from organisms like bacteria into our artificial technology, aiming to make the process more efficient than nature itself. One day it could create a range of materials, from chemical fuels to plastics.

What have wasps ever done for us?

→ **We hate them buzzing around at picnics, but wasps undoubtedly provide several key ecosystem services. Like their more-loved bee cousins, they are pollinators, and we simply couldn't live without them.**

Wasps suffer from a reputation problem, yet they are important pollinators, making them vital for agriculture and our food system. Also, as apex predators, they kill other insects, including pests, which protects our crops. Their venom and saliva have compounds that could become medicines.

Pollination is one of the most critical regulating services in the natural world. As pollinating insects and other animals feed, they move pollen from one plant to another and fertilise them. Three-quarters of our crops depend on pollinations, a service worth over $250 billion per year. Almost 90 per cent of all wild flowering plants need pollinating animals – at least in part. Wasps visit over 900 species of plant, some of which are entirely dependent on them to reproduce. Monkeys, rodents, birds and bats are all pollinators, too. This is a tremendous regulating service for humans, sustaining our global food system.

Crops, though, would be nothing without the soil they grow in. Soil is a free, supporting ecosystem service – providing the basis for all others. Soil is formed through the erosion of rocks, which contain all the elements plants need to grow. Diverse microbial communities and plants then colonise, and other organisms like earthworms move in to help cycle nutrients through the ecosystem, one which purifies air and water. Over time, the soil becomes a hotbed of organic matter and an important habitat for many species.

Biodiversity keeps soils healthy, and vice versa. Soils have provided us with an enormous variety of plant species to consume. Eating diverse foods is critical to our nutrition. Our ability to grow food has improved consistently throughout our history too, through increasingly intensified agriculture. Yet nature's ability to deliver the sustainable, diverse food we need is weakening. Our farming practices, and our global food system more generally, is the leading cause of biodiversity decline. Boosting biodiversity throughout our food system will help to preserve our nutrition overall.

AGROFORESTRY

One of the most promising farming techniques to emerge over recent decades is agroforestry. This practice aims to move our food production away from the harmful and often unproductive systems of monoculture. In agroforestry, trees, shrubs, grasslands and animals are integrated with growing crops in a way that fosters sustainable generation of food, and improves soil quality. Numerous studies have shown that agroforestry increases biodiversity in agricultural areas, locks nutrients within the ecosystem and can improve crop yields.

How do other species protect my health?

⟶ **Biodiversity helps keep our minds and bodies in good shape. Healthy ecosystems lock in diseases and may even prevent antimicrobial resistance, whereby our drugs are becoming ineffective against wily microorganisms such as bacteria and viruses.**

Healthy ecosystems and the resulting biodiversity provide a support system for our well-being. The relationship is complex and multifaceted, and research is still in its infancy – there is a lot of correlation, less so causation.

Some environmental factors are more obvious and fairly straightforward: air free of pollutants and clean fresh water are clearly good for human health. Trees also provide shade, helping us to regulate our temperature. These positive health effects are particularly beneficial in urban areas, where the quality of the environment is generally worse.

Other environmental factors are less obvious. Allergies are rising dramatically around the world, and researchers are questioning whether biodiversity loss is the cause. In 2013, scientists proposed that the disruption and replacement of our microbe symbionts could be the underlying factor. This 'biodiversity hypothesis' also suggested that the macroscale biodiversity – plants, animals and so on – is directly linked to the diversity at the microbial level. And that our increasing urbanisation could be detaching us from microbes in the environment that we need.

As for the link between biodiversity and mental health, the research is growing. Several studies have shown that spending time in green spaces is beneficial for our well-being. One recent paper found that our mental health improves in tandem with higher species richness – the number of species of plants and animals within an ecosystem.

Healthy ecosystems also lock in disease. Diseases that jump from animals to humans are called zoonoses, and there are many out there. But research shows higher biodiversity prevents both the transmission of such diseases within an ecosystem and their release from it. More species diversity means a pathogen may run into a less hospitable host, or a worse transmitter. This suppression of disease is known as the 'dilution effect'; its reverse, seemingly driven by our actions, is the 'amplification effect'. Scientists are now exploring whether microbial biodiversity in soil and freshwater ecosystems also works like an ecological barrier, preventing the spread of antimicrobial resistance.

ONE HEALTH

The health of people, plants, animals and our shared environment are interdependent and closely linked. Human activity, including trade, travel and land use, can heavily influence the health of ecosystems and therefore the spread of disease. In 2003, following epidemics of SARS and avian influenza, the concept of 'One Health' emerged, which aims to unify approaches to stabilise the health of all organisms and our environment. Despite the recent impetus, the idea can be traced back over two centuries.

How do we get aspirin from a tree?

→ **From the bark of a willow! Humans have been using plants as therapeutics for thousands of years. After turning to synthetic alternatives, pharma is once more looking to nature for new medicine.**

Humans have self-medicated for millenia. Neanderthal teeth, 40,000 years old, suggest they consumed aspirin and penicillin in plants and mould. A clay slab over 5,000 years old lists Sumerian drug recipes, using mandrake, poppy and over 250 other plants. Ancient Maya used resins from trees to prevent tooth decay.

As scientific methods improved, we developed ways to extract medicinal qualities from natural products and turn them into powerful medicines. The bark of the willow tree contains salicin, a chemical with pain-relieving qualities. The German pharmacologist Johann Andreas Buchner (1783–1852) isolated and named it in 1828. After some chemical tinkering, the pharma company Bayer sold the first over-the-counter aspirin tablets in 1915.

Such medicines have dramatically impacted human history. They helped us survive disease and extend our lifetimes. Quinine, an antimalarial compound found in the *Cinchona* tree, enabled tropical exploration – as well as European colonialism. Medicines derived from nature will no doubt influence our future.

Where to search for new drugs? Ecosystems with high species diversity are natural laboratories for the emergence, spread and prevention of disease. As such, they are models for biomedical research.

We have sourced medicines from the centre of rainforests and the depths of the ocean. Extracts from sea squirts led to a new anti-cancer drug. Bacteria living in symbiosis with nematodes produce promising antibiotics. There are antibacterial wound-healing compounds in Komodo dragon blood. Scientists are scouring deep-sea sediments, from the Arctic to Antarctica, in search of new therapeutics. Extremophiles, which deal with immense pressures, are a popular target.

Genetic engineering is another option: scientists have already tweaked the tobacco plant to produce the beginnings of a chemotherapy drug. Advances in genomics let scientists screen for therapeutic compounds at ever faster speeds. A few decades ago, pharmaceutical companies stopped looking to nature as they found new drug-making efficiencies in labs. Now the focus has once again shifted, back to the natural world.

ZOOPHARMACOGNOSY

Many animals, terrestrial and marine, have been observed seemingly taking advantage of natural medicines. This concept is called zoopharmacognosy. Chimps apparently chew leaves to cure parasites. Birds disturb ants so they get sprayed with an acid, which rids them of blood-sucking lice. Black tamarins, ocelots and deer all use the bark of one tree in Brazil's Atlantic forest, which has antifungal properties. Proving causation is tricky, but if humans self-medicate, why not other animals, too?

How did the tourist save the gorilla?

→ **Admittedly not without its issues, eco-tourism can boost awareness and help conservation, from financing to monitoring the poaching of animals such as Rwanda's beloved mountain gorillas. Done right, it is a win-win for all concerned.**

Biodiversity is a major draw for human visitors. Tourism can bring problems, from the disturbance of natural ecosystems to the polluting effects of travel, but by encouraging an interest in the natural world, it can also give a huge boost to biodiversity.

The ecosystem conditions that promote cultural good and benefits for humans are known as cultural services. These can be tangible or intangible, and range from recreational activities, to aesthetic appreciation, to artistic inspiration, to spiritual reflection. These services are grounded in the cognitive and physical interactions we have with nature, connections that probably deepen in line with ecosystem complexity. Cultural services, more than any other, inspire us to value nature for its intrinsic value, rather than as a resource for our own gain.

Cultural ecosystem services can be difficult to quantify, as they are based on subjective experience. But it's clear that over millions of years, the mutual interactions between humans and nature have shaped us both.

Ecologically-minded tourism offers a direct, practical gain. Tourists benefit from safaris through their appreciation of animal and plant life. This also pays for more rangers out in the field, who monitor endangered animals and prevent poaching. Through citizen science projects, amateur conservationists can use apps to track and monitor wildlife, providing valuable data for scientists. And tourists can help restore ecosystems: for example, by spearfishing invasive lionfish that have spread from North America to Brazil, decimating other life along the way.

By visiting national parks to see protected mountain gorillas, tourists generate over $100 million per year for Rwanda's government. This money flows back into the conservation of natural spaces and so boosts overall biodiversity. Mountain gorillas were once in a perilous state of decline, but now they are the only species of great ape whose populations are on the rise – a rare conservation success.

PROTECTED AREAS

In a sense, ecotourism is a mutually beneficial symbiotic relationship between humans and nature. People gain non-material benefits, and their money helps to regenerate the ecosystems they visit. One strategy to maintain this link is through protected areas: clearly defined regions defended by laws or other means and managed in order to preserve nature. There are nearly 300,000 protected areas around the world, in the oceans and on land. Ecotourism is a growing and sustainable way to support them.

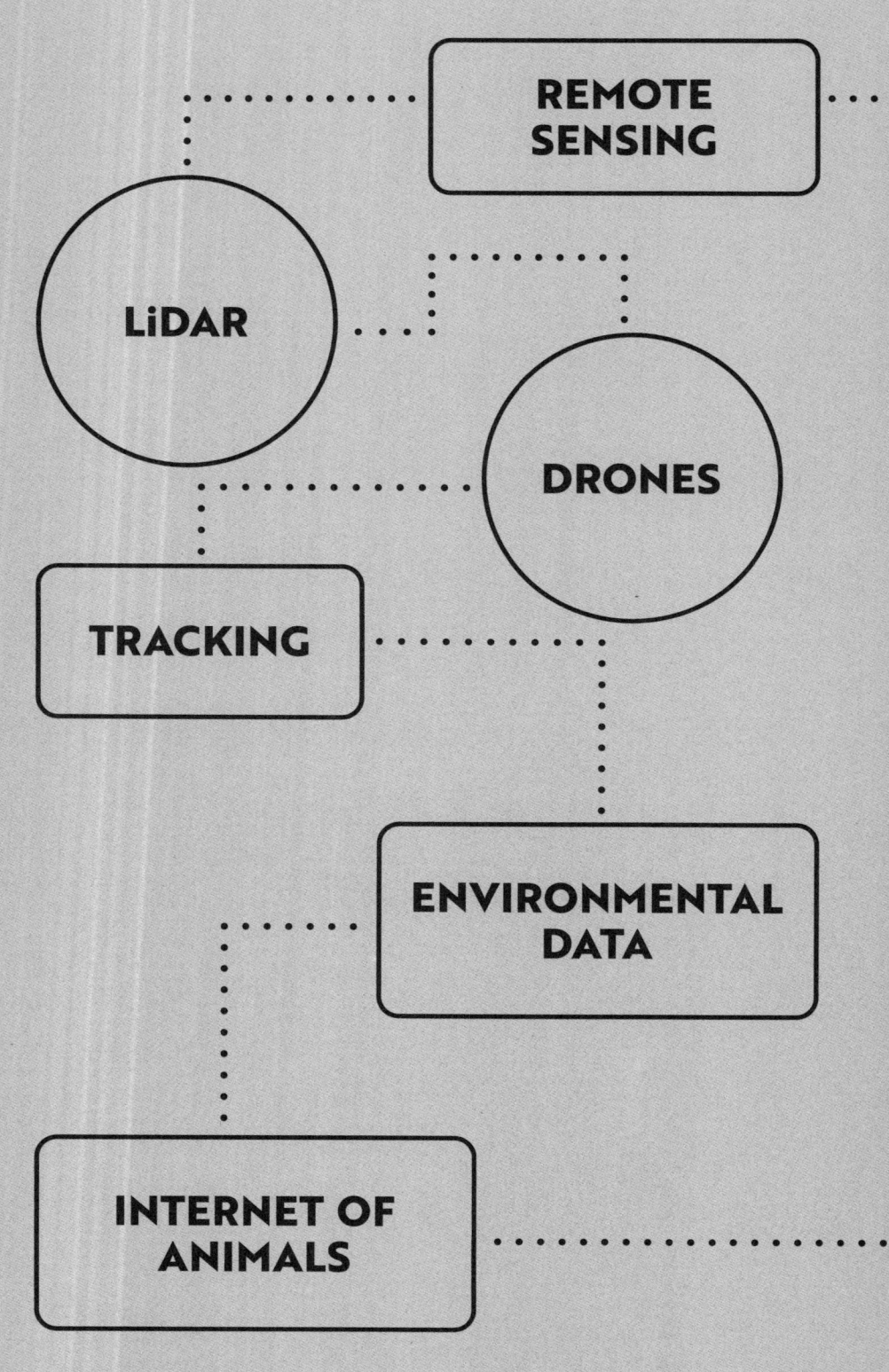

CHAPTER 5
TECHNOLOGY

- ECOLOGY
- ARTIFICIAL INTELLIGENCE
- ROBOTS

INTRODUCTION

Understanding the biosphere is key to protecting it. Knowing where species live, breed, interact and move helps us interpret changes in species traits – behaviours or attributes of species – which can impact ecosystem services. Gathering **ESSENTIAL BIODIVERSITY VARIABLES (EBVS)** such as community composition or population abundance can also indicate the health of habitats, and inform and assess national and international targets on biodiversity conservation.

Scientists collect data on biodiversity using many different techniques, ranging from good old-fashioned eyesight to advanced technologies for spotting, surveying and analysing an organism's health, activity and behaviour. All these methods have pros and cons, and examine nature at different qualitative and quantitative scales. In some parts of the world, records from amateur and professional scientists stretch back hundreds of years, revealing shifts in biodiversity over space and time. At its most basic, the data shows either the presence or absence of a species – though a lack of the former doesn't necessarily mean the latter.

Ecologists can use this data to create a **SPECIES DISTRIBUTION MODEL**, a popular framework that can feed global impact assessments of the past and future. These combine global species observations with environmental variables. Expanding networks of advanced satellites are illuminating environmental drivers with greater levels of resolution and accuracy than ever before, from the atmosphere to the ocean.

Raw species data comes from many sources, which differ in their scope and quality. These include natural history collections, manual surveys, global databases like the Global Biodiversity Information Facility (GBIF), national park inventories and expert assessments. There has been a large growth in the availability of such data, though ecological and geographical biases still exist.

Technology is helping us construct a far more accurate assessment of the world's biodiversity, how it is changing and what may come next – including in places human eyes cannot reach. Scientists invented the first **HYDROPHONES** around the First World War, to detect icebergs and submarines; now we are using them to listen to marine life around the world. Radio transmitters appeared in the 1950s, followed by **SATELLITE TELEMETRY** in the 1990s. **CAMERA TRAPS**, first used in 1877, have transformed conservation by giving us a peek at the secret behaviour of wildlife – and many hilarious selfies.

A **GEOPHONE** can pick up elephant behaviour from vibrations in the ground. Thermal imaging and sonar help us spot animals in the metaphorical dark and the actual dark. Vast networks of stations and sensors relay data around the planet. Advanced computing and machine learning can analyse data at blistering speeds. This is in high demand: technological advances have outpaced analytical techniques for some time, leading to a backlog in data.

We are still limited in our ability to identify the complexity and diversity of life on Earth, but with new tools to unravel the past and predict the future of species, the global picture is becoming a little less blurry.

TECHNOLOGY MAP

MONITORING

DATA

INTERNET OF ANIMALS (IOA)
Large-scale infrastructure combining continuous monitoring, datasets and artificial intelligence to track animals in near-real time.

SATELLITE TELEMETRY
Allows researchers to track animal movements using orbiting satellites that detect signals from a transmitter attached to the animal.

CAMERA TRAP
Special camera for providing data on the location and behaviour of species, population sizes and how different species interact with one another.

ENVIRONMENTAL DNA (EDNA)
The DNA left behind by plants and animals in an environment, which can be collected and sequenced by scientists to monitor biodiversity.

REMOTE SENSING TECHNOLOGY
Powerful cameras and sensors on aircraft and satellites used to gather data from Earth's surface about the environment and ecology.

TRACKING

ESSENTIAL BIODIVERSITY VARIABLES (EBVS)
Six core variables, such as community composition and population abundance, identified by scientists in order to focus research and analysis on key aspects of biodiversity.

GEOPHONE
Acoustic device that responds to ground vibrations and can be used to track the movements of large mammals such as elephants.

HYDROPHONE
Microphone specially designed for use underwater to record or listen to sound.

SPECIES DISTRIBUTION MODEL
System that uses environmental data to assess and predict species distribution geographically and over time.

ACOUSTIC LURES
Broadcasts of the sounds made by animals used to stimulate recolonisation of a habitat, with the scope to one day rebuild whole communities.

SOUNDSCAPE ECOLOGY
Technology used to monitor biological, geological and man-made sounds across landscapes and seascapes to reveal changes in the natural world.

GLOBAL POSITIONING SYSTEM (GPS)
Highly accurate global network of satellites and receiving devices for determining the position of a place or object on Earth.

DOPPLER SENSOR
Instrument that aims a constant signal of low-energy electromagnetic radiation at a target area, then analyses the signal reflected back.

LiDAR TECHNOLOGY
Use of lasers to scan the environment from the ground and air to create 3D models of the landscape, which can determine habitat quality.

Can you spot an elephant from space?

→ Yes! Scientists use a variety of technologies to analyse species and ecosystems from a distance, including sensors on the satellites orbiting Earth.

Scientists have long sought to understand and explain the diversity of species spread across our planet. Knowing why certain areas are particularly rich in species can help us to study biodiversity decline and decide where to invest limited conservation resources.

Over the past few decades, remote sensing technology has become a pivotal tool for conservation. Researchers now have access to the many powerful cameras and sensors on aircraft and satellites that harvest data on life below. These can capture high-resolution images of the ground, or detect electromagnetic energy from the Sun that is reflected or emitted from the Earth's surface. This reveals information about the underlying environment and ecology.

Conservationists can use remote sensing to measure the biophysical properties of plant life and environmental variables. As these properties shape the distribution and abundance of species across landscapes, this can help us to understand how and why organisms live where they do.

Researchers can map and monitor habitats; observe invasive species and illegal poaching; spot wildfires; and assess the health of entire ecosystems such as rainforests and coral reefs.

LiDAR technology is also widely employed in biodiversity research. This uses lasers to scan the environment from the ground and air, creating 3D models of the terrain and vegetation growing on it – key determinants of habitat quality.

Drones are another significant tool for monitoring and tracking wildlife – often with the help of AI-based detection methods – and these can also be equipped with remote sensing technology. Marine scientists use drones to map the ocean, measure a wide range of environmental variables and observe underwater creatures.

With so many satellites looking down on us, including the growing array of low-orbit satellites, scientists and conservationists can build up an ever more accurate understanding of life on our planet.

REMOTE SENSING AREAS

Remote technologies have become critical tools in biodiversity research and conservation. Hundreds of satellites now monitor our planet's ecosystems. With advances in remote sensing, scientists can distinguish species through direct observation, to the level of individual animals, trees and plants. They really can spot an elephant from space.

How do birds sleep when they fly?

→ Some do it using just half their brain. We know this because wearable technology is letting us dive even deeper into the complex behaviours of animals, and even understand how ecosystems are changing.

During the Second Punic War (218–201 BCE), a Roman officer tied a thread to the leg of a swallow, which flew to a besieged garrison and delivered the message it carried – the first reported use of bird banding. In 1902, the American zoologist Paul Bartsch (1871–1960) attached numbered bands onto black-crowned night herons, introducing a scientific banding system still used today. Bartsch probably would be amazed at how animal tracking has progressed.

Over recent decades, advanced wearable tracking technology has become smarter, cheaper and smaller. Today's scientists track animals as they migrate across continents and dive to the depths of the oceans. The arrival of high-tech, low-cost and miniaturised devices – along with the data they generate – has ushered in a 'Golden Age' of animal tracking.

Scientists can capture animal movements, unique behaviours and even internal states. We know great white sharks travel through ocean vortices to feast in the mesopelagic zone, where light fades to black. A sea turtle armed with a video camera recorded its encounter with a predatory tiger shark: the turtle bit and lunged at it before fleeing unscathed. Frigate birds, which soar over the oceans for months on end, sleep using half their brain as they fly.

Behaviour is usually the first thing to change in an animal following a variation in its environment. By combining animal tracking with fine-scale environmental data, we are starting to unpick the generative drivers behind these behaviours, too. Animals are revealing the health of habitats and intricate shifts within ecosystems – and the potential role climatic factors play. They directly monitor the environment, and reveal it too: sharks pointed us to the largest seagrass ecosystem in the world; deep-diving seals just showed us a hidden canyon in east Antarctica – just under a kilometre deeper than the supposed ocean floor.

Technology will advance further and become less invasive. Scientists are developing tattoo-like tags and sensors built into flexible material skins. Others just recorded the brain activity of a freely moving octopus – not an easy feat. Perhaps one day these cephalopods can tell us what they're thinking about.

INTERNET OF ANIMALS

Scientists are now building a giant animal monitoring infrastructure known as the Internet of Animals (IoA). The IoA will combine continuous monitoring, vast datasets and artificial intelligence to allow anyone to track animals in near-real time: from individual insects to flocks of migrating birds. It will answer many questions in conservation and ecology, help us spot poaching and disease, and could even give us earlier warnings about earthquakes and volcanoes.

If a tree falls in the woods, is the sound useful to scientists?

→ Perhaps, but the sonic hole it would leave in the ecosystem certainly would be. Soundscape ecologists monitor ecosystems to tune in to the sounds of the natural world, from birdsong to the unforgettable call of whales.

The dawn chorus is an unmistakable outburst of birdsong at daybreak. Clearly, more songs equals more birds. Now scientists can draw far more from this avian ensemble, from the community of species present even to the health and structure of the ecosystem itself.

It's not just birds having their vocals checked. From the forests of Borneo to the soils of the Arctic, scientists working in the field of soundscape ecology are tuning in to the remarkable diversity of sounds produced in the natural world.

Ecosystems are loud places. Animals in rainforests and deserts whoop, chirp, growl and roar to communicate. Coral reefs, once thought to be silent, are an orchestra of clicks and pops. Even plants make noises when stressed, though they are out of the range of human ears.

With ever-better audio equipment to both record and analyse sounds, we can tease apart the entire acoustic diversity of soundscapes. From the bottom up, we can detect individuals from their unique signals. This can help scientists measure biodiversity and uncover rare or even cryptic species – those that superficially seem identical to a different species altogether. From the top down, the acoustic fingerprint of an ecosystem – and how it changes over time – can tell us about overall community health, which species are entering and which have departed.

Soundscape ecology is an efficient, cost-effective way to monitor the status and flow of biodiversity across broad landscapes and seascapes. Soundscapes include biological, geological and man-made sounds, and these can all reveal how the living world is changing – and what the drivers may be. A fallen tree will remove parts of the audio, and so reflect the health of the habitat.

Now the field is advancing from passive monitoring to active restoration. Acoustic lures – broadcasts of sounds created by animals – have been used to stimulate recolonisation in animals as diverse as frogs, bats and whales. This technique could one day be used to rebuild entire communities.

WHALE SONG

Sound is not a new tool for conservation. The haunting majesty of whale song is now well known. Singing may have even saved their lives. The American scientist Roger Payne (1935–2023) discovered that whales sing to communicate with each other in 1967. He used the recordings to kick-start a conservation movement to protect whales, which had been vanishing from our oceans – mostly because of our harpoons. Payne's work helped to galvanise global public pressure, which led to the ban of commercial whaling in 1986.

If a fish committed murder, could we tell which one did it?

→ Yes, although you'd still need a good enough detective. But thanks to the advent of environmental DNA, we'd be able to tell which species were there at the time.

Just as humans leave traces of DNA at a crime scene, so too do all plants and animals. Recent advances in the collection and sequencing of this environmental DNA (eDNA) is revolutionising the monitoring and conservation of biodiversity.

No animal need be captured, or even still be at the scene. Marine animals like fish litter the sea with their DNA as they swim along; the fur or scat (faeces) of a snow leopard will leave a genetic marker; plant roots, pollen, eggs, skin, saliva, urine: DNA is everywhere.

Scientists extract DNA from environmental samples like soil or water, and then amplify segments of interest in a lab. Using a technique known as DNA barcoding, they can then find individual species – even the shy ones. eDNA combined with metabarcoding – a way of quickly analysing multiple DNA samples at once – can give fast and accurate assessments of biodiversity in ecosystems. It's an efficient way to spot invasive species early, like the destructive American bullfrog that is currently hopping through more than 40 countries. It's particularly useful in scanning bodies of water, peering into the microbial world, and searching for emerging pathogens.

DNA can survive for a long time, too, which lets us paint complex portraits of former worlds. In 2022, scientists uncovered a once-thriving 2-million-year-old ecosystem in present-day Greenland using eDNA. They found signatures of trees like birch and poplar, and Arctic shrubs and herbs – many of which were new to the area – along with animals such as hares, mastodons and reindeers. Evidence of horseshoe crabs and algae pointed to a climate far warmer than today's.

Marine biologists have used eDNA to measure biodiversity as it shifts through layers of the ocean, and analyse community differences in ports. Others found tardigrades (microscopic invertebrates), Tibetan snowcocks and domestic dogs on the peak of Mount Everest. The technique is relatively new, but already evolving. Scientists have now developed a way of sucking eDNA straight out of the air, making the process even more efficient. Shotgun sequencing, a molecular technique, will soon make it cheaper and easier, too.

HUMAN GENETIC BY-CATCH

Scientists recently showed that eDNA sampling techniques easily detect a lot of human DNA, something they call 'human genetic by-catch'. This has raised certain ethical questions. Who does the DNA belong to if it's scattered through the environment? How should or could this data be protected? There are opportunities from this discovery too, like detecting humans involved in the illegal wildlife trade or finding sites of archaeological importance.

When is a jellyfish no longer a jellyfish?

→ When it's a robot. Advances in robotics have led to everything from hardworking tree-planting robots to robotic jellyfish that clean coral reefs, giving conservationists new tools to monitor and protect biodiversity.

Robots of all shapes and sizes are already working to preserve biodiversity across the world. One major application is monitoring. Drones have been used to collect eDNA from tree branches. There are many autonomous and semi-autonomous drones underwater – equipped with things like computer vision; Global Positioning Systems (GPS), pressure and doppler sensors to align themselves; and others to measure environmental data – drifting along with sea creatures, keeping an eye on them and their ocean home.

Echo-sounding robots are uncovering biodiversity hotspots in hard-to-reach habitats of the deep sea. Flying robots soak up the snot of passing whales, and give this valuable data back to scientists. Researchers at the Max Planck Institute recently created a robot that looks and swims like a jellyfish. The pumping swim stirs up currents below, which real jellyfish use to bring in nutrients. The robotic one does it to pick up fragments of plastic, and could be used to delicately clean up coral reefs.

Robots can influence biodiversity, too. There are robots that plant trees, far faster than a human could. Scientists in Germany are developing robotic honeybee hives, complete with systems to monitor bee health and even to alter their behaviour. Robot dogs are scampering through the forests outside of Pisa, checking up on the trees.

Robots can also kill, and so remove, harmful life. Australian scientists designed submersible robots that inject fatal bile salts into the crown-of-thorns starfish, a poisonous invasive species that feeds on coral reefs.

And they can harvest, too. A robot raspberry helps train other robots to pick fruit. One day robots could work in precision fishing, reducing by-catch.

While there is much optimism around robotics and conservation, some caution that robots are not a panacea. They risk exacerbating biodiversity decline – for example, through the processes and resources needed for their production, and the pollution from their disposal. Perhaps more than anything, the risk is that more robots may lessen the perceived human responsibility to protect the natural world.

ROBOT SWARMS

Humans have taken much inspiration from nature to design new robots. This includes their movement, too. The collective behaviours of certain animals – schools of fish, herds of bison, flocks of birds – follow simple rules that scientists are recreating in machines. Roboticists are working to develop swarms of unmanned drones, intelligently communicating with each other and acting in unison to deliver specific tasks. Robot swarms could clean up oil spills, pollinate crops and sample biodiversity en masse.

Will AI save the world's biodiversity?

→ It's certainly already a powerful conservation tool. From spotting migrating salmon to illegal goldmines, artificial intelligence is giving us a deeper understanding of biodiversity, and its decline.

AI is already having profound impacts on scientific research, including on the climate, biodiversity and conservation. At its core, machine learning uses mathematical algorithms to analyse vast datasets and produce predictions or decisions. Most of the technological developments detailed in this chapter either rely on or produce huge streams of data. Humans may take hours, days or weeks to analyse a dataset, while an algorithm could do it in seconds.

This can speed up some of the vital decisions we need to take to protect biodiversity. One recently designed AI program named CAPTAIN considers the costs and benefits of prioritising certain conservation areas. The system takes in data on biodiversity, climate and even conservation budgets. It can simulate ecosystems like coral reefs, or rainforests, and predict the outcomes on biodiversity under a range of potential environmental and anthropogenic factors.

AI is helping scientists better understand biodiversity, too. Recent advances in computer vision – a subset of machine learning – means that images from all over the biosphere can be quickly interpreted. Researchers can track salmon populations in Alaska using sonar imaging. Previously, a human had to count every fish, but now a machine has taken over. AI can classify species within images, or films, at astounding speeds and with ever greater accuracy.

Programs can be trained to learn and spot behaviours in animals, which can help classify and monitor individual species. At the broader level, AI can figure out species diversity within a community, by parsing through huge tranches of genetic data or analysing complex soundscapes. It's widely used to analyse environmental variables relevant to biodiversity. Algorithms can predict wind speeds from videos of blowing trees and evaluate how walrus populations are adapting to a changing Arctic. And it is also a powerful tool for measuring biodiversity decline, and the drivers behind it – counting illegal goldmines in the Amazon rainforest, for example.

CITIZEN SCIENCE

AI is indirectly helping conservation by engaging citizen science. Machine learning algorithms are behind several nature applications that identify species in images or from the sounds they produce. These apps let us collect and even analyse data on biodiversity pretty much anywhere in the world. This helps scientific research projects in a cost-effective way, providing much-needed information across space and time. It also increases public engagement in science. The more people connect with nature, the more they care.

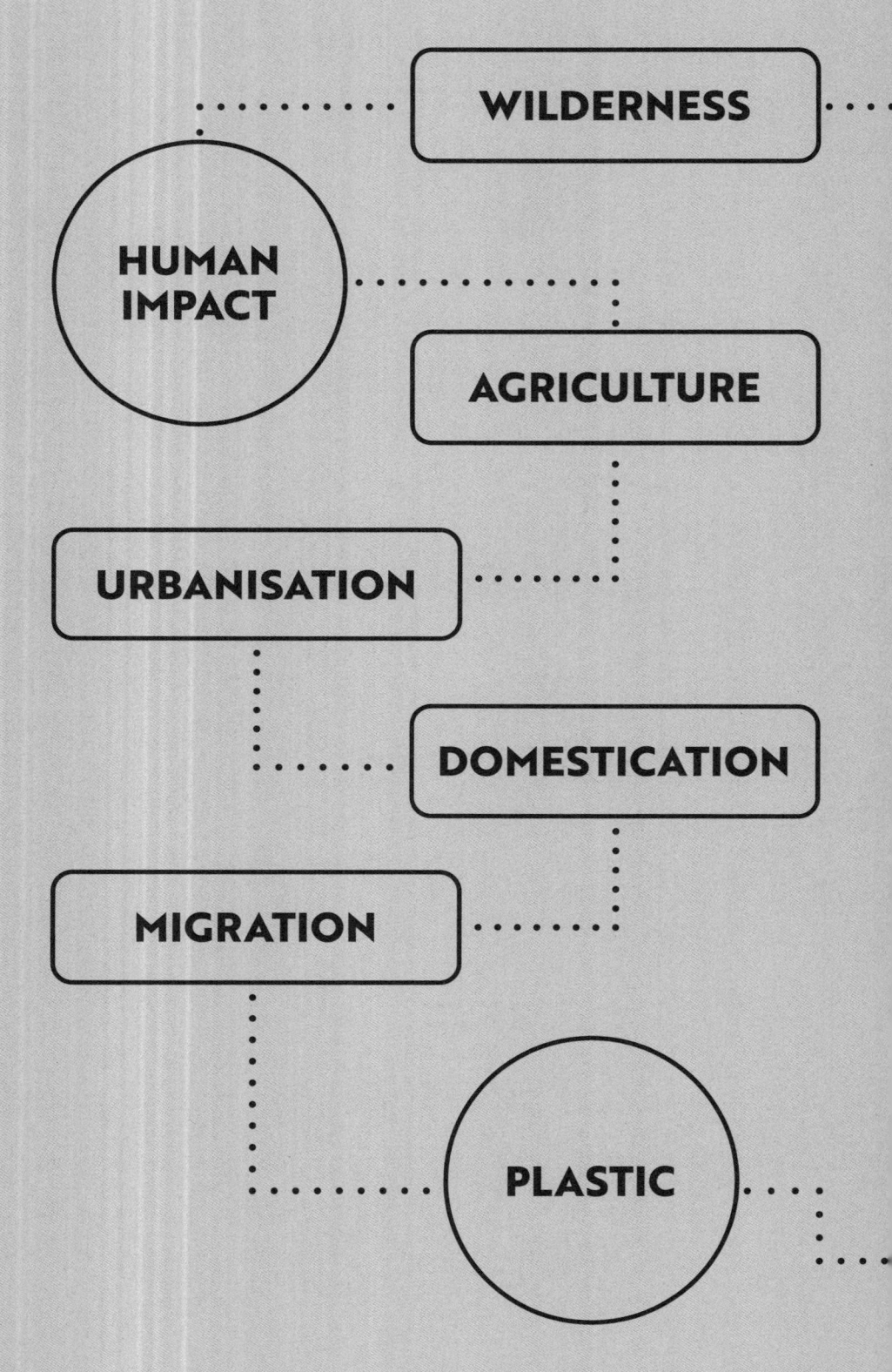

CHAPTER 6
THE ANTHROPOCENE

INVASIVE SPECIES

EXTINCTION

INTRODUCTION

Imagine Earth's 4.6-billion-year history was one 24-hour day. The first life appeared around 4am. At midday, the atmosphere became rich in oxygen. Plants colonised land at 9:30pm and animals followed at around 10pm. Humans arrived a fraction of a second before midnight.

In reality, the Earth's geological past is divided into eons, eras, periods, epochs and ages. These units of the **GEOLOGIC TIME SCALE** are marked by major events in deep history, those which significantly alter the make-up of organisms on the planet – an example is the asteroid that wiped out most dinosaurs and ended the Mesozoic era (252 to 66 million years ago).

Officially we are still in the **HOLOCENE**, an epoch that began after the last ice age around 12,000 years ago. Humans have existed for a geological eye-blink, yet we have had such profound impacts on the physics, chemistry and biology of our planet that some scientists suggest we have entered a new epoch, the **ANTHROPOCENE**.

If the Anthropocene were formally declared a geologic epoch, this would essentially mean that human activity had dislodged the planet from the natural planetary systems that previously governed the Earth. It would mean humans have such dominance over the planet – through the use of land, fertiliser, fossil fuels, and so on – that we rival other geological and tectonic forces.

For example, our long-term climate – including the onset and end of ice ages – is thought to be strongly driven by **MILANKOVITCH CYCLES**. These cycles relate to the tilting and wobbling of the Earth as it orbits the Sun, as well as the shape of the orbit itself. Recent research suggests the burning of fossil fuels has delayed the onset of the next ice age, and may delay it for another 100,000 years. As the climate shapes biodiversity, our impacts could fundamentally change the life that exists in the future.

Assigning a new geological division is not trivial. The International Commission on Stratigraphy (ICS) must declare it, based on evidence found in layers of rock known as strata. The main indicator is fossils, which point to the existence and loss of life. The marking must be significant in oceans and continents, and last for many thousands or millions of years into the future. There are several proposals for delineating evidence in our time, including atmospheric chemistry, plastics, concrete and the mass destruction of wildlife.

The definition could encourage progress on lowering carbon emissions and tackling biodiversity loss, or reclassify our relationship with the natural world: a biome could be termed an **ANTHROME**, for example, and managed in a way that acknowledges our deep-rooted influence on global ecosystems. In 2024, the ICS rejected the Anthropocene, though another vote remains possible. Meanwhile, many within the scientific world will continue to use the term in recognition of our short-lived, yet transformative, impact on the planet.

THE ANTHROPOCENE MAP

CHANGE

MIGRATION BLOCKER
Human infrastructure, border or barrier (such as a wall or fence) that prevents the migration of animals such as birds, bats and butterflies.

ANTHROME
Anthropogenic biome that occurs as a result of sustained interactions between humans and their environment.

ANTHROPOCENE
Unofficial geological epoch when human activity began to substantially alter Earth's surface, atmosphere and oceans. Some scientists suggest this period began around 1950.

GYRE
Ocean vortex due to circulating currents formed by wind patterns. There are six main vortices in the world's oceans, which now contain plastic litter such as microplastics and plastic bottles.

GOLDEN SPIKE
Global marker in the rock record that reveals changes in fossils or minerals as a result of a global event, which lasts through time.

LANDFILL
Method for disposing of large amounts of rubbish by burying it in deep holes at designated sites.

HOLOCENE
Geological epoch that began after the last ice age, around 12,000 years ago – we are still officially in the Holocene.

EPOCHS

GEOLOGIC TIME SCALE
'Calendar' revealing Earth's geologic history and divided into time units of descending duration, including an eon (a billion years) and an epoch (several million years).

MILANKOVITCH CYCLES
Cycles including Earth's tilt and the trajectory around the Sun that combine to affect long-term climate patterns on our planet.

BACKGROUND EXTINCTION RATE
The rate at which species become extinct due to natural evolutionary flow, estimated at one species per million each year.

HYBRIDISATION
Creation of hybrids as a result of cross-breeding between different species

BIOTIC HOMOGENISATION
Process by which the huge diversity of plants and animals is gradually replaced by a smaller group of plants and animals.

DOMESTICATION
Gradual process of bringing wild plants and animals under human control to ensure a supply of food (crops and meat) and labour (animal power).

FERALISATION
Reverse process to domestication in which domesticated animals are accidentally or intentionally reintegrated into the wild.

HUMANS

Where can I find wilderness?

→ **Unfortunately, there's not much left, such has been the profound impact of humans on the planet over thousands of years – and in particular the last few hundred. Our species has put its mark on most of the natural world.**

↳ There is no absolute consensus on the definition of wilderness. In general, areas considered wild are mostly free from the impacts of human pressures (though not necessarily humans). In these relatively undisturbed regions, environmental, ecological and evolutionary processes continue without human interference.

Roughly 30 million square kilometres (12 million square miles) of our land – just under a quarter – is still wilderness. Since the early 1990s, around 3.3 million square kilometres (1.3 million square miles) has been lost, a patch larger than India. Ocean wilderness has dwindled to 13 per cent. Most wild marine space is confined to polar regions, and very little remains by the coast. Even in remote parts of the Amazon, prehistoric peoples changed the plant composition of the forest through early farming and domestication – albeit in sustainable ways.

Human expansion is driving the decline of the wild. We have scraped sediment from waterways, and our dams have prevented it from entering others. Our industrialisation, mechanised agriculture, land and sea use change, destruction of habitats, overexploitation of resources, burning of fossil fuels and the resulting climate change are all fundamentally transforming natural areas. Our growing population will push up the need for resources, which risks further loss of wilderness.

This activity surged following the Second World War. The period from the 1950s onward is known as 'The Great Acceleration', a step-change in global human industrialisation intertwined with technological progress and environmental change. This formidable expansion of population, energy use and greenhouse gas emissions may have destabilised cyclical features of the world shaped through deep time. Some propose this should mark the beginning of the Anthropocene.

Wilderness areas are vital. They act as refuges for biodiversity: evidence suggests species are half as likely to go extinct in wild areas. And they help to maintain and stabilise the Earth's systems that foster life on local and global scales, so it stands to reason we should endeavour to preserve and protect them as much as possible.

GOLDEN SPIKE

Our impact on the world and its wilderness is obvious. Nevertheless, scientists are still debating whether or not we are officially in the Anthropocene. Geological ages are defined by global markers in the rock record known as 'Golden Spikes', such as changes in fossils or minerals. They must last through time. A good example of our impact is the first nuclear test in 1945. The testing of nuclear weapons littered the planet with radioactive isotopes that will be detected for millions of years.

Are cities good or bad for biodiversity?

→ **Both. Removing habitats to build cities has been destructive without a doubt. But urban areas can harbour biodiversity too, and could even be part of the solution.**

⇶ One of the most profound and evident transformations of the natural landscape is cities. Our expanding population and technological advances – notably concrete and steel – have created giant urban habitats in a stark break from the natural world.

Soils are now covered in concrete. Vast structures of glass and metal tower into the skies. We have destroyed, removed or covered habitats and drained resources from the surrounding lands. We are ecological engineers, though mostly for the benefit of our own species. Cities are entirely new ecosystems, which merge the natural and man-made worlds.

Cities were long thought empty of life – the 'biological deserts fallacy' – which isn't true. Yet they have certainly altered it. Urban areas tend to favour generalist species that can adapt to a range of habitats, like foxes, squirrels and pigeons. Some species avoid cities altogether, while others choose to live in them. Cities can even affect evolution: an analysis in 160 cities around the world showed that white clover plants produce fewer anti-herbivore defence chemicals the closer they are to city centres. Consider it urban chic.

Cities are hostile environments, and threaten biodiversity. Yet they are also home to many endangered species, including some – like the yellow-flowered San Francisco lessingia – which may now only be found in urban areas. Some city-dwelling species are more productive, grow faster and in larger populations than outside. Urban birds can have larger brains; mammals larger litters.

Urbanisation isn't going away. Cities must therefore become part of the solution, and many are now incorporating nature-based solutions into their future planning, harmonising artificial ecosystems and helping biodiversity.

Whether they will remain in the geological record is up for debate. Evidence of our roads, offices, aqueducts and other infrastructure will likely last for some time: the construction of cities has altered rates of erosion and sedimentation the world over. But even cities recently abandoned, lost and rediscovered by humans have begun to crumble back into the soils.

URBAN DIET

City-dwelling wildlife can feast on a variety of new food, from hamburgers, to nasi goreng, to just about anything found in our garbage, including faeces. Urban human gut microbiomes differ from those outside cities, and food could be a reason. Some urban animals (coyotes, sparrows and lizards) share more similar gut microbiota to urban humans than their rural kin – and similar to animals living in distant cities – suggesting urbanisation is the driver.

How did the chicken take over the world?

⟶ **By getting in with the humans. We have shaped the world's biodiversity to suit our needs, to the benefit of some species and the demise of others.**

Nearly 13,000 years ago, the red junglefowl lived wild, in the trees of tropical South-east Asia. At some point, perhaps around 1,500 BCE, it was brought into human society as a domestic bird. Now known commonly as the chicken, there are over 33 billion of them, living pretty much everywhere around the world. Future geologists analysing fossils from today would find evidence of an extreme blending of biodiversity within our brief time on the planet.

The exact processes and timings of domestication are still debated. So too is the definition, though the broad idea is that one species has an influence over the reproduction and care of another, to increase a certain resource. Also, that both species benefit in some way from the relationship. (It's not just humans, by the way: Agriculturally-minded ants started cultivating a handy fungus millions of years ago, and some fish in Belize have domesticated shrimp to fertilise their algae farms.)

What is clear is that humans have filtered biodiversity on a mass scale to serve our niche: spreading around the planet those species – or certain traits – that are beneficial to us and getting rid of those we deem unnecessary or harmful. It's quite a change: humans and our livestock now account for 96 per cent of all mammals.

Domestication has allowed our species and civilisation to advance. Early farming and the tending of animals may have fostered a sedentary lifestyle that enabled our populations to grow. Domestication of plants and animals may have crossed over: evidence suggests rice agriculture drew the junglefowl down from the trees into our homes. Certain crops like rice, wheat and millet have fuelled our expansion around the world – and theirs. Domesticated horses helped us travel further, and faster. Domesticated dogs gave us companions and protectors along the way.

Domestication changes the underlying genetics of animals. The brains of most domesticated animals are smaller than those of their wild counterparts, an effect that correlates with increased human contact. And the promotion of just a few plants and animals has come at the demise of many wild species, including pollinators and crops, which are in stark decline around the world.

FERALISATION

Feralisation is domestication's lesser-known twin. Feral animals are domesticated ones that re-enter the wild – either intentionally or by accident. This reintegration, and the hybridisation that follows, can disrupt ecosystems and is generally bad news for biodiversity.

Feral animals may have genes less adapted to the wild, which they can spread; and they can outbreed their wild cousins. Feralisation increases the extinction risk of wild species. And like the swine trouncing around North America, feral animals can cause a lot of damage.

What if humans never invented sailing or flight?

→ **Life would still spread around the planet, as it always has. But our technologically advanced modes of exploration and travel have undoubtedly ramped up the globalisation of biodiversity.**

For life, travel to distant lands is nothing new. As new islands break through the ocean surface, they are steadily colonised by a succession of beings. Plant seeds migrate via ocean currents. Insects hitch a ride on atmospheric currents of wind. Even larger animals like monkeys can be carried on natural rafts that float over the oceans. Yet no prior migration of organisms rivals the introduction of non-native species caused by humans. For millennia we have deliberately carried alien species for food, trade or sport. Many spread unintentionally as stowaways.

Boats provide ample surfaces for all kinds of biotic life to latch on to, from barnacles, to tubeworms, to bacteria. Air travel has helped spread organisms into harder-to-reach areas of land. Cars and trains, and their underlying infrastructure, provide easy corridors through which organisms can spread. Even undersea cables offer a nice hard substrate along which to disperse.

The extent to which today's snapshot of biodiversity is due to prior human influence is difficult to say. Scientists have found evidence of a large influx of plant species into the UK during the Late Bronze Age. Trade networks thousands of years old linking the Middle East with Africa, Europe and South Asia may have scrambled species over much of the planet.

Human-induced dispersal has also caused a large rise in invasive species. These alien organisms often have no natural predator, thrive in a new environment and can decimate native species. Not all human introductions are harmful, though: think of the potato, originally cultivated by the Incas in the Andes and now a staple on many dinner plates. Some species introductions are even beneficial.

Biological invasions have increased dramatically over the past two centuries, and in particular over recent decades, coinciding with mass human migration, technological advances and the Industrial Revolution. As technology improves, and new routes are opened across land, sea and air, the number of alien invasions are likely to keep ticking up.

MIGRATION BLOCKERS

Migration is one of the natural wonders of the world. Thanks to advances in science and technology, we now know that far more animals migrate than previously thought. However, our infrastructure, borders and barriers all hamper migration – blocking it completely, distracting animals with artificial lights, or killing them along the route. (Estimates suggest hundreds of millions of migrating birds die each year in the USA alone by colliding with buildings.) This can have knock-on effects in ecosystems and even cause the loss of some species altogether. While we're shipping some organisms around the world, we're unintentionally halting others.

Am I eating plastic right now?

⟶ It's possible. Research suggests all kinds of organisms are eating or absorbing plastics – even up to a credit card's worth a week for humans. But the long-term effects of this new material still aren't clear.

Plastic was designed to stand the test of time, and could mark the beginning of the Anthropocene. This hardy material has helped us enormously. Plastic packaging keeps food and other goods fresh for longer. Plastics are sterile and used medically for items such as syringes, surgical gloves and bandages. They are in cars, trains, planes and satellites. Plastic toys entertain children. Plastic condoms prevent them, too.

A side effect of our addiction to plastic is that we have lavishly distributed it all over the planet. Plastic doesn't biodegrade, so it often gets shoved into landfills as waste. Much of it ends up in the sea, or other waterways. Some plastic pieces are manufactured to be tiny – for facial scrubs, say. Larger pieces can also break down to form microplastics, which have spread everywhere across land and sea.

Microplastics have been found near the peak of Mount Everest, in deep ocean trenches and in Arctic ice. In Antarctica, it snows plastic. Pieces of plastic have been extracted from human lungs and found in our hearts. It's very likely that plastics will form a layer in the rocks, a colourful yet depressing indication that we are living in a 'plastics age'.

The extent to which plastic damages biodiversity isn't clear, largely because not enough time has passed for us to know for sure. Larger pieces choke, trap or maim animals, often leading to slow and painful deaths. Microplastics accumulating in fish harm their growth, and even change their behaviour. Research has linked plastics with an uptick in coral-killing diseases. They can absorb harmful pollutants from the water, too. Hermit crabs keep using plastics for their homes, and are dying inside en masse.

As plastics enter marine life, it's easy to see how they travel up the food chain to us. Plastics are in soils and enter plant roots. One recent study suggests we eat up to 52,000 plastic particles per year. It's still not clear if all this plastic is a risk to animal health, but it's probably not great.

GARBAGE PATCHES

The currents carrying water, heat and nutrients around our oceans also transport plastic litter. Microplastics, along with larger pieces like bottles and fishing gear, have accumulated in six major ocean vortices known as gyres. The largest, the Great Pacific Garbage Patch, is roughly 1.6 million square kilometres (618,000 square miles) – three times the size of France. Although this ecosystem seems to be thriving, with a diverse range of life, from sea snails to the Portuguese Man-of-War, the patches are a giant hazard for other marine life.

Were there once lions in Europe?

→ Yes – and in Asia – but you won't hear them roar: these populations died out long ago. Now wild lions are confined to Africa, and their range is shrinking. Like many species, they could soon become extinct.

⇉ Perhaps the most prominent signal left behind by our age will be the fossils marking a catastrophic and rapid loss of biodiversity. The weight of evidence suggests we are in the early stages of the sixth mass extinction in Earth's history, one triggered not by natural causes but human activity.

Species only exist for a while before they become extinct, and others take their place. This natural evolutionary flow is calculated as the background extinction rate, thought to be around one species per million each year. Mass extinctions happen when the rate of loss far exceeds that of replacement. Scientists estimate the current extinction rate to be between 100 and 1,000 times the background rate, the fastest in human history – though this is complicated by the fact that many species haven't even been identified yet.

Extinction comes more easily to certain groups, including rare and endemic species, and top predators. Amphibians seem the most at risk today. Even species that aren't threatened with extinction are facing widespread population collapse. Liveable ranges for organisms are shrinking far faster than normal, choking them out of existence in certain parts of the world. Since 1970, wildlife populations have plummeted 69 per cent on average.

This kind of population-level extinction is normally a precursor to species-level extinction. Wild lions, for example, once prowled Asia, Europe and the Middle East but now only live in Africa – where their range is shrinking sharply.

As the oceans absorb more carbon dioxide, they could become so acidified that corals aren't able to build reef structures. In the fossil record, this would appear as a 'reef gap', something that has occurred in each of the past global extinctions.

Over time, rare species adapted to unique conditions will be replaced by those capable of tolerating the world as we are shaping it. The startling diversity of the natural world will be steadily replaced by a smaller set of plants and animals, resulting in biotic homogenisation – another likely hallmark of the Anthropocene.

MASS EXTINCTIONS

Mass extinctions are classified as such when roughly 75 per cent of life on the planet is lost within 2.8 million years. There have been five previous mass extinctions. Fossil boundaries often separate geological periods, as evidence of transitional events in Earth's history. The Cretaceous–Tertiary mass extinction, 66 million years ago, wiped out 78 per cent of species, including most of the dinosaurs – except those that became birds. This was likely triggered by a massive asteroid striking present-day Mexico, helped by massive volcanism in India.

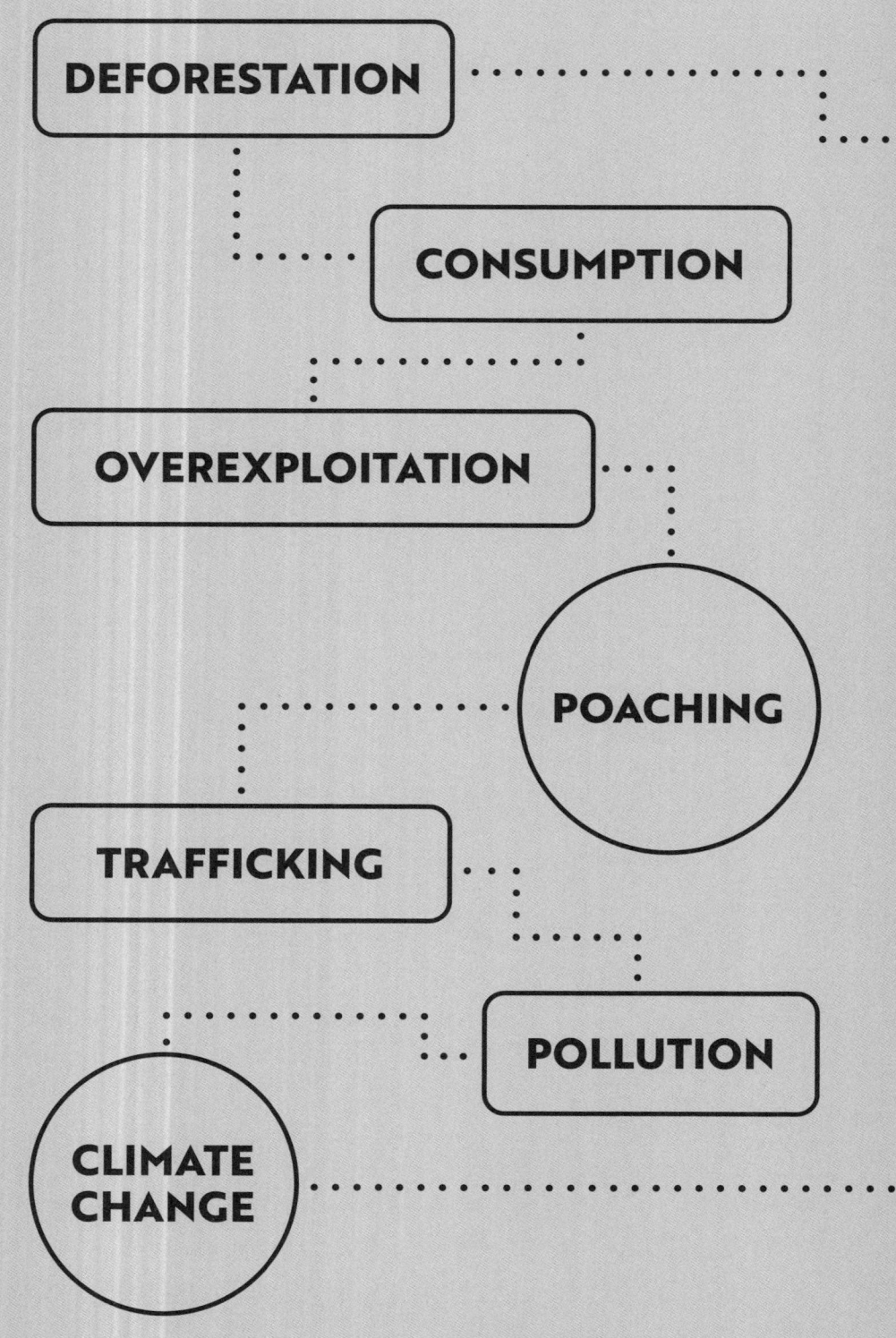

CHAPTER 7
LIFE UNDER THREAT

- LAND USE
- INVASIVE SPECIES
- EXTREME WEATHER EVENTS
- PESTICIDES

INTRODUCTION

The loss of any species is tragic. But as ecosystems are interconnected, it can also have cascading rippling effects with unpredictable consequences. Systemic loss presents broader problems for both nature and ourselves.

One million plants and animals are threatened with **EXTINCTION**, many within decades. This figure is conservative and could be vastly underestimated, as we still do not know how many species exist. Extinction rates are accelerating. Plants avoided most of the previous mass extinctions, yet they are threatened in this one too. A global review in 2019 found that 40 per cent of insect species could become extinct over the next few decades. Insects pollinate our crops, control plant pests and help to recycle nutrients. If they disappeared completely, our food system would collapse, along with that of most birds and amphibians.

Extinctions can be global or local (**EXTIRPATION**). The main causes are a loss of habitat and genetic diversity. The dramatic rates of current biodiversity loss are likely the result of the compounding influences of humans upon nature. The main drivers of decline are dynamic, and local context plays an important role – islands are more susceptible to **INVASIVE SPECIES**, for example.

A combination of these drivers can cause an **EXTINCTION VORTEX**. This model suggests that a combination of environmental stress, shrinking populations and lowered genetic diversity weakens communities of organisms. Catastrophic events, such as a huge storm or excessive human hunting, swirl the vortex further and make reproduction and survival impossible.

Over 20 per cent of the world's ecosystems are at risk of collapsing or shifting into another state. Research suggests this could happen sooner than anticipated this century. Ecosystems may be hit with destructive feedback loops of local drivers like species decline combined with extreme events such as wildfires and broader patterns of climate change. The collapse of one ecosystem could then have a domino effect on others, a phenomenon harrowingly known as **ECOLOGICAL DOOM-LOOPS**, leading to a rapid global destabilisation in the not too distant future.

What does all this mean for the future of biodiversity? One probable scenario is that life will become less unique. Specialist organisms are likely to be lost and replaced by cosmopolitan and generalist species that can live alongside humans and in a variety of habitats. Ecological communities will become less distinct, a process known as biotic homogenisation. Some scientists have proposed a term for this widespread simplification of nature: the **HOMOGECENE**.

However, while on global scales biodiversity appears to be in precipitous decline, evidence suggests that on local scales the same may not be true. Research indicates that in some places species richness – the number of species within an ecosystem – is stable, and may even be rising. Some islands have many more species than they did in the past. Invasions are leading to new hybrid species, which may never have existed without human influence. Nature finds ways to persist, but we can't assume that evolution will preserve the life that sustains our civilisation.

LIFE UNDER THREAT MAP

EFFECTS

ALGAL BLOOM
Rapid increase in the population of microscopic algae in a body of water; although natural, blooms can be harmful to ecosystems.

POLYCHLORINATED BIPHENYL (PCB)
Harmful chemical used in industrial and consumer products such as pesticides. Although banned, PCBs are still produced and are found in the oceans.

EXTINCTION
Loss of a species of animal or plant on the death of all remaining living members.

QUATERNARY EXTINCTION
Mass die-off of predominantly larger mammals such as the woolly mammoth between 50,000 and 5,000 years ago, largely driven by human hunting.

ECOLOGICAL DOOM-LOOPS
Knock-on, domino-like effect of ecosystem collapse caused by the transformation or collapse of a neighbouring ecosystem.

EXTINCTION VORTEX
Downward spiral in a small population due to environmental stress, shrinking populations and a loss of genetic variability. Often exacerbated by catastrophic events such as storms.

TRAGEDY OF THE COMMONS
Describes a vicious cycle in which humans will always use more resources for themselves to the detriment of the wider public good because of inherent selfishness.

RESOURCES

THREATS

DEAD ZONES
Ocean areas that are hypoxic, meaning they no longer have enough oxygen to support marine life.

EXTIRPATION
Local eradication of a species, such as a bird forced into extinction by the loss of a rainforest through logging.

HOMOGECENE
Theoretical future era that will come about due to the increasing homogenisation of species across the globe as native flora and fauna are replaced by invasive non-native species.

BY-CATCH
Animals caught by the fishing industry that the fishermen do not want, cannot sell or are not allowed to keep by law.

INTERCROPPING
Agricultural practice of growing two or more crops together rather than cultivating a monoculture.

INVASIVE SPECIES
Living organisms that invade and inhabit areas beyond their natural range, harming other life and the environment.

BIOSECURITY
Prevention of disease-causing agents from entering or leaving a location where they might pose a risk to humans, animals or food quality.

BUSHMEAT
Meat from wild animals. A primary source of protein in tropical forested areas of Africa, Latin America and Asia; an important food resource and commodity for poor communities.

Do young Florida panthers move far from their parents?

→ It can be hundreds of miles, but now that their habitat has been chopped down to make way for agriculture, they are restricted to one small corner of the Sunshine State. Our land use has a massive impact on most species.

The Florida panther once thrived across several states, from Arkansas to South Carolina. But as most of its habitat has been lost, the remaining 200 or so live in a patch of land measuring less than 10,000 square kilometres (4,000 square miles). Juveniles disperse over long distances to set up their own range, but now pickings are slim.

We have transformed ecosystems across at least three-quarters of ice-free land. As we change land to meet our needs, we destroy and degrade habitats for other life. Our activities include logging, mining, building cities and the pollution of water and soil. Land use – specifically our global food system – is the leading cause of biodiversity decline.

Roughly 80 per cent of threatened birds and mammals are so imperilled because of agriculture. Our demand for meat, particularly beef and lamb, requires a lot of land, either for grazing or growing crops to feed animals. New farmland is created at the expense of natural habitats; most species can't survive in these new landscapes.

Between 1980 and 2000, 100 million hectares (250 million acres) of forest were lost, largely for cattle ranching in Latin America and the expansion of palm oil plantations in Southeast Asia. One study estimates that 15 billion trees are felled each year. We have lost 46 per cent of our forests since the last ice age.

Future population expansion threatens species extinctions, largely in the tropics, where land use change is likely to accelerate. As incomes rise, our demands change, placing further strains on biodiversity. Consumption – for food but also energy, fashion and technology – increases in parallel with income. Global calorie intake rose 31 per cent between 1961 and 2013. Wealthier humans demand more meat and therefore more land.

Changing our consumption patterns can make a real difference. So can improving crop yields, through improved management or intercropping plants. Where land clearing happens, maintaining connections between isolated habitats will slow extinction rates and make ecosystems more productive too.

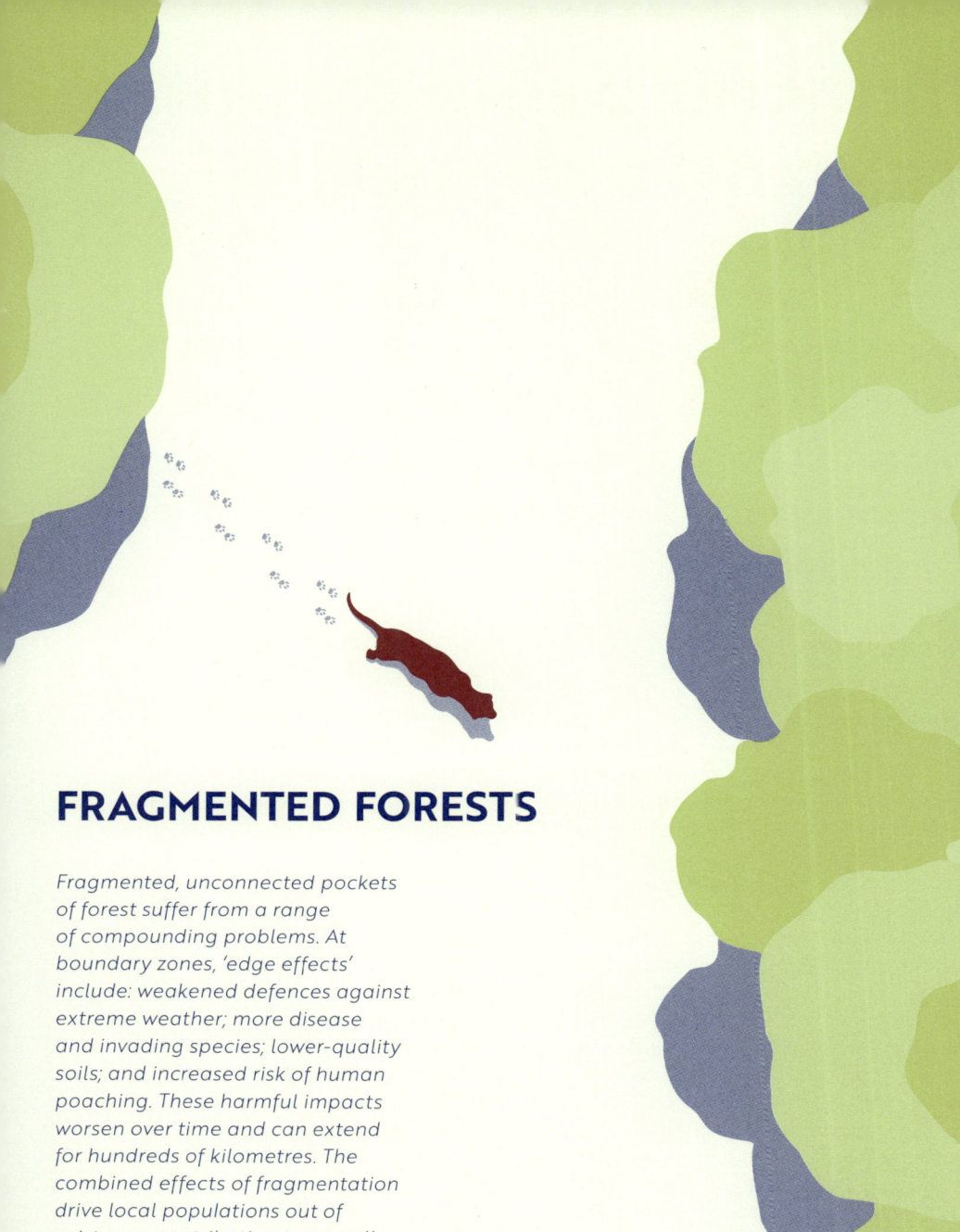

FRAGMENTED FORESTS

Fragmented, unconnected pockets of forest suffer from a range of compounding problems. At boundary zones, 'edge effects' include: weakened defences against extreme weather; more disease and invading species; lower-quality soils; and increased risk of human poaching. These harmful impacts worsen over time and can extend for hundreds of kilometres. The combined effects of fragmentation drive local populations out of existence, contributing to overall extinction and significantly harming the functioning of ecosystems. Connectivity is key.

Why do people want to traffic pangolins?

⟶ To use their bodies in traditional medicine. The demand for animals and their parts fuels a global illegal wildlife trade worth billions of dollars. This is just one driver behind the human overexploitation of biodiversity.

⤳ After land use change, the second largest contributor to biodiversity decline is our exploitation of wild things. For thousands of years humans have hunted, gathered and fished. We have chopped trees for timber, drained fresh water and mined the land. In many cases, we've done this to unsustainable levels. Seventy-two per cent of species listed as threatened or near-threatened are overexploited for our gain.

Humans have a long history of overexploiting: we are, at the least, partly responsible for the Quaternary Extinction, a mass die-off between 50,000 and 5,000 years ago that eradicated mostly larger animals, including the woolly mammoth, the sabre-tooth tiger and the Siberian unicorn. Hunting for body parts and bushmeat still threatens many species today, particularly large herbivores.

Fishing has the most impact on marine biodiversity. Industrial fleets allow us to fish further and deeper than ever before. Over a third of fish stocks are overfished and commercial fishing vessels often seek top predators like tuna, which can trigger trophic cascades. By-catch – the unintended capture of marine species – makes up 40 per cent of the global haul, often killing threatened animals like sharks and turtles. Illegal exploitation compounds these issues. Illegal, unreported and unregulated (IUU) fishing accounts for one in five caught fish, a major and global problem that hampers conservation efforts.

More broadly, the illegal wildlife trade is valued at up to $10 billion annually and presents a grave threat to biodiversity – one fuelled by the rise of the internet and social media. Pangolins are hunted for their flesh and scales, which are used in a broth thought to have medicinal qualities. A million pangolins have been poached and trafficked illegally over the past decade, driving them towards extinction in the wild.

There are solutions to this over-exploitation. These include greater enforcement of hunting regulations, efforts to stem IUU fishing across national borders and on the high seas, international agreements against animal trade, and greater education on the source of things we consume, both illicit and legal.

GOVERNING THE COMMONS

In 1968, the ecologist Garrett Hardin (1915–2003) suggested humans are inherently selfish and use more resources for their own benefit to the detriment of the wider public good, a vicious cycle known as 'The Tragedy of the Commons'. A more optimistic solution was proposed by Nobel-winning political scientist Elinor Ostrom (1933–2012). In her book, Governing the Commons, she showed how many local communities have developed successful common resource strategies, without the need for public or private intervention.

Can corals keep up with climate change?

→ It's possible some in the deep ocean will survive, though most corals seem likely to die over the next century. Climate change is already driving many organisms to the brink of their survivable ranges.

Corals are often referred to as the rainforests of the sea. They are complex colonies of animals, made up of many individual reef-building invertebrates called polyps. Coral reefs sustain vibrant marine communities and have provided food, environmental stability and culture for human societies for millenia. But around the world, they are vanishing.

Climate change is the main culprit. Acidic oceans harm coral growth. Melting ice caps are raising sea levels, increasing damaging sedimentation. And corals are incredibly sensitive to thermal stress. During marine heatwaves, which are becoming more frequent and severe due to climate change, corals expel their symbiotic algae. This has led to several mass bleaching events, causing widespread coral death. If global temperatures rise to 1.5°C (35°F) above pre-Industrial levels, 70–90 per cent of corals will die by 2050. If we reach 2°C (36°F), 99 per cent of corals will disappear.

Climate change is already affecting the physiology, genetics and behaviour of many plants and animals. The ones to survive will be those that can tolerate climatic changes, move to new ranges, or adapt. Many animals are seeking out cooler climes, either by moving to polar regions or climbing higher up mountains. Endemic species in the tropics, whose survival ranges are more rigid, are most at risk. Other species will probably benefit from their new environments.

Extreme weather events such as flooding, wildfires and storms are expected to become more intense and happen more frequently. Australia's 2019 bushfire season killed or displaced roughly 3 billion animals. Those that didn't die saw their habitats swept away in flames.

The changing climate is expected to redefine the functioning of ecosystems in the sea and on land, including the timing of migrations, reproduction and hibernation. Climate change has long been the third largest contributor to biodiversity decline, though it will probably move up to the top spot this century.

Climate change will play a major role in how biodiversity and ecosystem functioning will survive in the future. Novel ecological communities, where species will co-occur in unknown combinations, are expected to emerge – with unknown ecological and environmental outcomes. Many plants will benefit: they love atmospheric carbon, though they will suck more water from the land as they grow.

ESCALATOR TO EXTINCTION

An organism can only go so far up a mountain before it runs out of room. As more species enter these alpine environments, competition is expected to rise and resources grow scarcer. In the short term, the number of species will increase, though over time many are expected to become extinct. Some already have. Mountain species have been described as being on an 'escalator to extinction'. Mountains in the tropics seem to be the speediest in this regard.

Why are Mexicans making bricks out of algae?

→ **When life gives you excess algae, build stuff with it. This is just one proactive approach to dealing with the destructive effects of pollution, which can exacerbate natural phenomena such as seaweed explosions, or algal blooms.**

Every year since 2011, Mexico's crystalline waters and pristine beaches have been clogged in foul-smelling brown seaweed. This algae forms part of the Great Atlantic Sargassum Belt, an annual bloom that stretches from North America to western Africa. It's visible from space and certainly from the shoreline.

Algal blooms are natural and can provide a vital habitat for many marine creatures, from eels to turtles to birds. But when fuelled to great sizes they can become deadly. Many algae produce toxins that poison waters.

The Great Atlantic Sargassum Belt, like other algal blooms around the world, is thought to be fuelled by the nutrients humans are adding to freshwater and coastal systems. Pollution, in the air, water and soil, is a large contributor to biodiversity decline – particularly excess nutrients, mostly nitrogen and phosphorus.

Fertiliser runoff from farming is a major culprit. Harmful blooms are also common around bays and coastlines near human settlements, where sewage leaches into the water. More than 80 per cent of wastewater around the world feeds into water systems untreated. Millions of tonnes of industrial waste enter water, including heavy metals that can disrupt ecosystem functioning.

Air pollution also harms organic life. Emissions of sulphur dioxide and nitrogen oxide are causing acid rain. Ground-level ozone, from cars or chemical plants, damages plant cell membranes.

Pesticides are toxic and harm other pollinators, like bees, and insectivorous birds. Insects and plant populations are falling due to the use of non-selective insecticides. Industrial chemicals called PCBs were banned in some countries nearly 50 years ago, but are still produced and persist in the oceans. These pernicious chemicals accumulate in food chains and threaten global orca populations with drastic decline.

The brick-making example shows that people can turn negatives into positives. In fact, seaweed is in popular demand, with startups creating fuel and food based on the stuff appearing around the planet. Where there is pollution, there can be innovation too.

DEAD ZONES

Dead zones are areas in the ocean that are starved of oxygen, largely caused by excess nitrogen pollution and harmful algal blooms. As algae grow, they consume oxygen and block out sunlight from plants below. When they die, they sink and are decomposed by bacteria, using up more oxygen in the water. This can devastate marine life, including fish, crabs and particularly bottom-living animals like shrimp. Scientists have identified over 400 dead zones around the world.

OCEAN

DEAD ZONE

COAST

Where is the lionfish going?

→ Anywhere it can. Invasive species like the lionfish are a major cause of biodiversity decline. With no predators and a voracious appetite, the lionfish is decimating wildlife around the world, and we are helping it.

Lionfish are beautiful – and deadly. Native to the warm waters of the Indo-Pacific, they are now so prolific they could come soon to a coastline near you. They were first detected off the Florida coast in the 1980s. Defended by poisonous spines and with no natural predator, they have munched their way through native species down the eastern coast of the Americas as far as Brazil. Lionfish have also invaded the Mediterranean sea, taking advantage of the Suez Canal.

Native species work their way into an ecosystem naturally. Non-native species are introduced by humans but do no noticeable harm. But invasive species inhabit areas beyond their native range and are harmful – to the environment or other life. They are normally generalist species, doing well in a range of environmental and ecological situations. Those that thrive in their new homes can outcompete and outbreed native animals. They can also bring new parasites and disease.

Invasive species are particularly harmful on islands, devastating native plant and animal life. One study linked biological invasions with around half the known extinctions on islands and as the main cause in places like Polynesia and Micronesia, New Zealand and the West Indies.

A landmark report in 2023 suggested the risks from invasive species have been underestimated. They play a key role in 60 per cent of all extinctions, and mitigating costs run at over $423 billion each year. And biological invasions are increasing. As we travel, we help spread alien organisms – so far over 37,000 species. Indirectly, human-caused climate change is also expanding the liveable ranges for invasive species. Invasive insects can travel long distances with wildfires humans are increasingly linked to. As the world's climate and systems change further, new pathways are likely to open up.

One benefit of lionfish is that they're a nutritious food, which humans are starting to enjoy (they taste great on the barbeque). Aside from eating them, we can also stop invasive species with improved biosecurity and import controls.

INVASIVE ALLIES

The cane toad is one of the most notorious invasive species. This little hopper was brought to Australia in 1935 to control pests in sugar cane plantations. They didn't help with the sugar cane pests, but became one themselves. Monitor lizards eat the toxic toads and die, removing a competitor of feral cats – another widely invasive species in Australia. This is boosting feral cat populations, creating an invasive trophic cascade that could lead to an 'invasional meltdown' of native species.

```
            IN-SITU
         CONSERVATION

    EX-SITU
  CONSERVATION

      SEED              CAPTIVE
      BANKS             BREEDING

   BIOBANKS

                   REWILDING

   DE-EXTINCTION
```

CHAPTER 8
CONSERVATION

- SYNTHETIC BIOLOGY
- INDIGENOUS PEOPLES
- GENE DRIVES

INTRODUCTION

The cultural roots of the modern conservation movement stretch back hundreds of years. Past philosophers, clergy, natural historians and laymen have all written of their wonder at the natural world and lamented the loss of its prior state. In 1569, hunting was prohibited on Karpfstock, a Swiss mountain, making it perhaps the first area set aside for conservation. In 1872, Yellowstone became the world's first **NATIONAL PARK**. While this idea was soon exported around the world, other communities and religions had been conserving lands for thousands of years.

The establishment of protected areas in the Western world, while not free from controversy, fed into a global conservation movement that developed through the 20th century. In 1948, the International Union for the Conservation of Nature (IUCN) was formed, a global network of government and civil society organisations that safeguards the natural world. Other national and international organisations have followed.

In September 1978, the American biologist Michael Soulé (1936–2020) spoke of the serious threats posed to human welfare by the rapid disappearance of plant and animal life. He described a 'biological diversity crisis that will reach a crescendo in the first half of the twenty-first century'. His pleas led to the emergence of **CONSERVATION BIOLOGY**, a multidisciplinary field aimed at understanding biodiversity and figuring out ways to preserve it. Its theories have been applied to the classification of **ENDANGERED SPECIES**, the design of **NATURE RESERVES** and **CAPTIVE BREEDING PROGRAMMES**.

A range of international policy frameworks have been set up to tackle biodiversity decline, such as the Convention on Biological Diversity in 1993 and the United Nation's Sustainable Development Goals in 2015. Yet many of the targets we have set ourselves have been missed, or likely will be.

So how should our lands and oceans be managed to benefit biodiversity and the climate, while sustaining expanding human populations? These are questions under consideration and debate. Should we try to coexist with species that we are trying to protect, or leave nature to heal itself while keeping human input at a minimum? Should we tinker with evolution through **GENETIC ENGINEERING**, and if so, when should we stop? Should we try to restore nature to some semblance of a former state, or recognise that humans have fundamentally altered almost every square mile of our planet and develop new ecosystems that acknowledge this?

In 2016, E.O. Wilson set out his proposal for 'Half-Earth': humans should leave 50 per cent of the planet for nature to recover, which would in turn save humanity and ensure a sustainable future. In 2023, a Half-Earth resolution entered into U.S. Congress, bringing Wilson's vision closer to reality. Our conservation actions have saved a number of species from extinction. We could also achieve a third of the necessary reductions in carbon emissions needed over the next decade by restoring nature, tackling the twin crises of climate and biodiversity loss in tandem. With concerted effort, we could again reshape the planet in a way that benefits all life.

CONSERVATION MAP

ACTION

NATIONAL PARK
Large area of land used for conservation purposes and appreciated for its natural beauty; managed and protected by national governments.

CONSERVATION BIOLOGY
Multi-disciplinary field that aims to understand biodiversity and develop ways to preserve it.

NATURE RESERVE
An area protected for its significant flora, fauna or other natural features, which plays a vital role in the conservation of animal populations and their habitats.

TRANSLOCATION
Movement of an organism to a location for the purposes of conservation.

REWILDING
Allowing a natural space to recover by giving it time to regenerate or by actively introducing species to reverse environmental damage.

ENDANGERED SPECIES
A species in danger of extinction in the wild, largely as a result of habitat loss and reduced genetic variation.

MEGAFAUNA
Large terrestrial animals such as elephants, rhinoceroses and large bovines. Animals such as bison are often used in rewilding programmes.

DE-EXTINCTION
Controversial long-term strategy for bringing animals such as the woolly mammoth back from extinction, for the benefits they could bring to the environment.

PRESERVATION

CAPTIVE BREEDING PROGRAMMES
The breeding and housing of endangered animals in man-made habitats with the potential for release into the wild.

INBREEDING DEPRESSION
The result of reduced genetic diversity, such as occurs in captive breeding programmes, which ultimately weakens a population's ability to reproduce and survive.

BIOBANK
Collection of animal tissues and molecular information, an example being the CryoArks project in the UK.

GENETIC ENGINEERING
Use of gene technology to alter the DNA of an organism to deliberately control and influence genetic traits.

SEED BANK
Type of gene bank where seeds are stored in order to protect genetic diversity for the future.

GENE DRIVE
Genetic elements that destabilise normal rules of inheritance, forcing a genotype through a population.

GENOME
Complete set of DNA instructions found in a cell. The human genome, for example, contains 23 pairs of chromosomes in the cell's nucleus.

GENOTYPE
Genetic material that makes up an organism and also a group of organisms with the same genetic composition.

SYNTHETIC BIOLOGY
Multi-disciplinary field that uses engineering principles and computational techniques to create artificial biological systems.

SCIENCE

How do we know when a species has been saved?

⟶ **When a species is stable or rising, it is considered 'saved'. You would hope that in this scenario the threats have declined and conservation measures have succeeded – but it could just be temporary.**

By the 1990s, the wild population of the Przewalski's gazelle, once spread across western China, had shrunk to less than 300. Conservationists marked it as Critically Endangered and enacted rescue programmes. Fast-forward to 2022 and there were over 2,800 of these wild antelopes bounding around Qinghai Lake.

Since 1993, conservation efforts have saved 48 birds and mammals from extinction. The extinction rate would have been four times higher without them. With conservation absent, the peregrine falcon, Spix's macaw, golden lion tamarin or the pygmy hog may have followed the dodo's doomed footsteps.

In-situ conservation aims to preserve biodiversity in natural habitats by, for example, monitoring and maintaining genetic variation within populations. Ex-situ conservation protects resources off-site and includes legal protections, storage of seeds or embryos and captive breeding programmes, where animals are housed in man-made habitats and potentially released back into the wild.

Captive breeding has pros and cons. It shelters animals or plants from threats and can augment or re-establish populations in the wild. But it is costly: on average over $200,000 per species per year. Many animals adapt to captivity, which alters their evolution and makes them less likely to survive in the wild. Captive breeding can stifle genetic diversity, leading to inbreeding depression, which weakens a population's ability to reproduce and survive. Breeding centres often share genetic resources to avoid this.

Reintroductions are, however, mostly unsuccessful. Scientists are now incorporating knowledge of the macro- and micro-worlds to explore whether altered gut and skin microbiota from different captivity conditions – such as diet or stress – could be a cause of this failure. Future conservation will also likely take the shifting dynamics of ecosystems under climate change into account.

Animals, fungi and plants are ranked for their extinction risk on databases such as the IUCN Red List of Threatened Species. Set up in 1964, it is updated by researchers around the world and lists over 150,000 species. When recoveries are reported, species can change category – away from extinction.

THE GREEN LIST

The IUCN recently established an optimistic counterpart to the Red List known as the Green Status of Species. This list highlights conservation success stories. Full recovery is assessed according to a species' presence and lack of extinction threat throughout its range; and whether it is ecologically functioning. Species receive a Green Score up to 100 per cent. This is ambitious and ambiguous by design, and helps to standardise recovery efforts across the biological kingdoms – up to now and into the future.

Can Noah's seed ark save us?

⟶ **It's almost certainly going to save a lot of seeds. Seed banks store genetic resources in case they are lost and need replacing. But seeds will only survive so long and must be replanted to ensure success.**

A giant store of the world's plant life lies buried deep inside a mountain in the snowy realms of the Svalbard archipelago, off the coast of Norway. The Svalbard Global Seed Vault, or 'Noah's ark of seeds', is one of over 1,750 seed banks hoping to preserve our plant life before it disappears. It acts as a backup repository for other facilities. Duplicates of over 1.2 million seed samples are kept in this remote location, safe from threats. Seed banks have helped re-establish plant life in Australia following devastating bushfires and restart collections lost during the Syrian civil war.

The erosion of plant diversity around the world is a major threat to our future food supply. Seed banking is a valuable tool for the preservation of plants and for translocation – the movement of an organism to a location for conservation. Scientists can use these resources to better understand plant growth and even breed new crop varieties, able to better withstand disease, floods and droughts, or provide greater nutritional value and even improve human health.

Seeds are dried and frozen and held in suspended animation to be regrown at future dates. Not all plants produce seeds, however, and other cryopreservation techniques are being used to preserve ferns and mosses, and algae in lichens. Scientists are also raising concerns that seeds may not grow if they are stored for too long.

Seed banks are not completely futureproof. Wild crop relatives of plants are disappearing along with valuable genetic information. And these institutions also face difficulties with funding around the world.

Biobanks of animal genetic information also exist, though are at an earlier stage of development. The CryoArks project in the United Kingdom is working to build a zoological bank of animal tissues and molecular information. The first pig biobank was established in Munich in 2015.

Biodiversity banks, along with assisted reproduction technologies, could play a larger and more active role in conservation in the future – by reintroducing genetic diversity, for example.

MAPPING EARTH'S GENOMES

Genomic initiatives complement biodiversity banks. The EarthBiogenome Project aims to sequence the genomes of every single living species on the planet. Genetic information provides a foundation for scientists and conservationists to understand biodiversity and how to conserve and restore it. These genetic libraries offer research avenues to prevent the spread of pathogens and boost ecosystem services. They could help us prevent further biodiversity decline in the face of climate change.

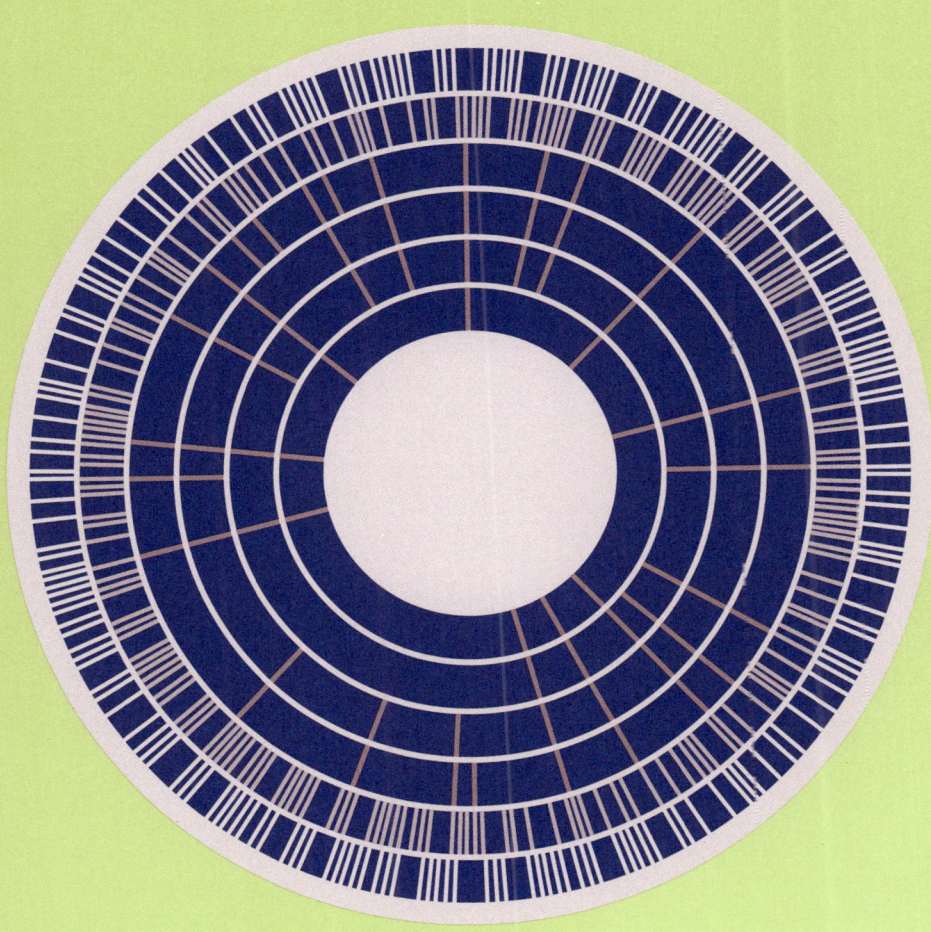

How many bison does it take to rebuild an ecosystem?

→ **Just a few! They instantly start thrashing around their environment, creating new ecological niches in which other organisms can thrive. Rewilding holds great promise in conservation.**

⇉ In July 2022, the woodlands around Canterbury received some old-time inhabitants: three female bison from Europe, the first wild ones to live in the UK for thousands of years. These huffing herbivores once roamed prehistoric forests, yet like many species became locally extinct. Ecologists brought them back to help regenerate degraded forests. By crushing through trees, urinating and defecating, bison create new habitats for birds, insects and other life. Later they were joined by other ancient breeds: Exmoor ponies and Iron Age pigs. Together they will build a wild wood.

Rewilding is a conservation idea over 30 years old which recently has picked up momentum. It aims to give nature a chance to recover a former state with minimal human input – either passively by creating space and time for it to do so, or actively by introducing certain species. The idea is that by returning megafauna, top-down 'trophic rewilding' will revive ecological systems, reverse environmental degradation and restore complexity.

Research into the ecological impacts of rewilding is still limited. Its proponents suggest it could halt biodiversity decline and create more resilient ecosystems. Critics point out there is no official definition of rewilding, or what 'wild' means, making overall goals difficult to envisage. The process is controversial, both in ecological circles and with the public, who sometimes cry foul at the reintroduction of potentially dangerous species.

There are many rewilding projects underway around the world with mixed success. Buffalo have returned to Yellowstone, as well as Alaska and Oklahoma. Giant tortoises reintroduced to an arid island in the Galapagos triggered a shift from savannah to grassland. Cheetahs returned to India, yet several died within months.

Some research suggests trophic rewilding could boost carbon sequestration through the creation of dynamic ecosystems. The recovery of blue wildebeest on the Serengeti transformed it from a carbon source to sink. In many cases, populations would have to reach 'ecologically meaningful' sizes. The success of rewilding may therefore depend on whether humans and wild animals can coexist across land and sea.

DE-EXTINCTION

One of the more ambitious long-term rewilding proposals is to bring back the woolly mammoth. This process, de-extinction, has never been done successfully. But several companies have been set up in recent years to achieve the goal, by using ancient DNA and breeding a mammoth hybrid in an elephant. Mammoths – and other large animals – could help transform the Arctic tundra and mitigate climate change. By turning trees into grassland, they could trample around, compacting the permafrost and preventing it from melting.

Should we gene edit mosquitos out of existence?

→ That's an open question, but we now feasibly have the technology to do so. Synthetic biology is allowing us to tinker with nature in extraordinary ways. But should we?

Genetic technology has advanced such that humans are now able to apply the principles of engineering to biological systems. This burgeoning field is known as synthetic biology. Synthetic biologists are able to circumvent nature and alter genes within organisms, giving them the ability to do things they may not normally be able to. For example, scientists have installed bioluminescent genes from jellyfish into plants, mice and marmoset monkeys, to make them glow.

Gene drives are genetic elements widely present in nature that subvert the normal rules of inheritance, actively spreading traits above the normal frequency by copying themselves or deleting other genes – forcing a genotype through a population. With synthetic biology, scientists are now able to create gene drives artificially. This opens up a range of possibilities for transforming biodiversity on our planet, and an equally large dose of ethical questions.

So far, synthetic gene drives have only been tested in the laboratory. But the advance of this technology over the past decade has been rapid enough that scientists can now fundamentally alter organisms and potentially even entire ecosystems by distorting inheritance.

Mosquitos are far more than a nuisance; they spread malaria among humans and lead to hundreds of thousands of deaths every year. Should we eradicate them? It's now feasible that we could. Artificial gene drives in mosquitos can make females sterile, for example. This trait spreads exponentially through generations of quickly breeding populations. One tweaked male released into the wild could potentially destroy his entire species.

The ethics are hotly debated. There could be unknown consequences of deleting a species. Gene drives could pass horizontally between species. But the tool could hold great potential in conservation, particularly as positive traits could be inherited too. Corals could resist warmer waters. Pathogens made less virulent. Invasive species could be eradicated. There is much potential in this technology. Reaching a consensus on its use may not be easy.

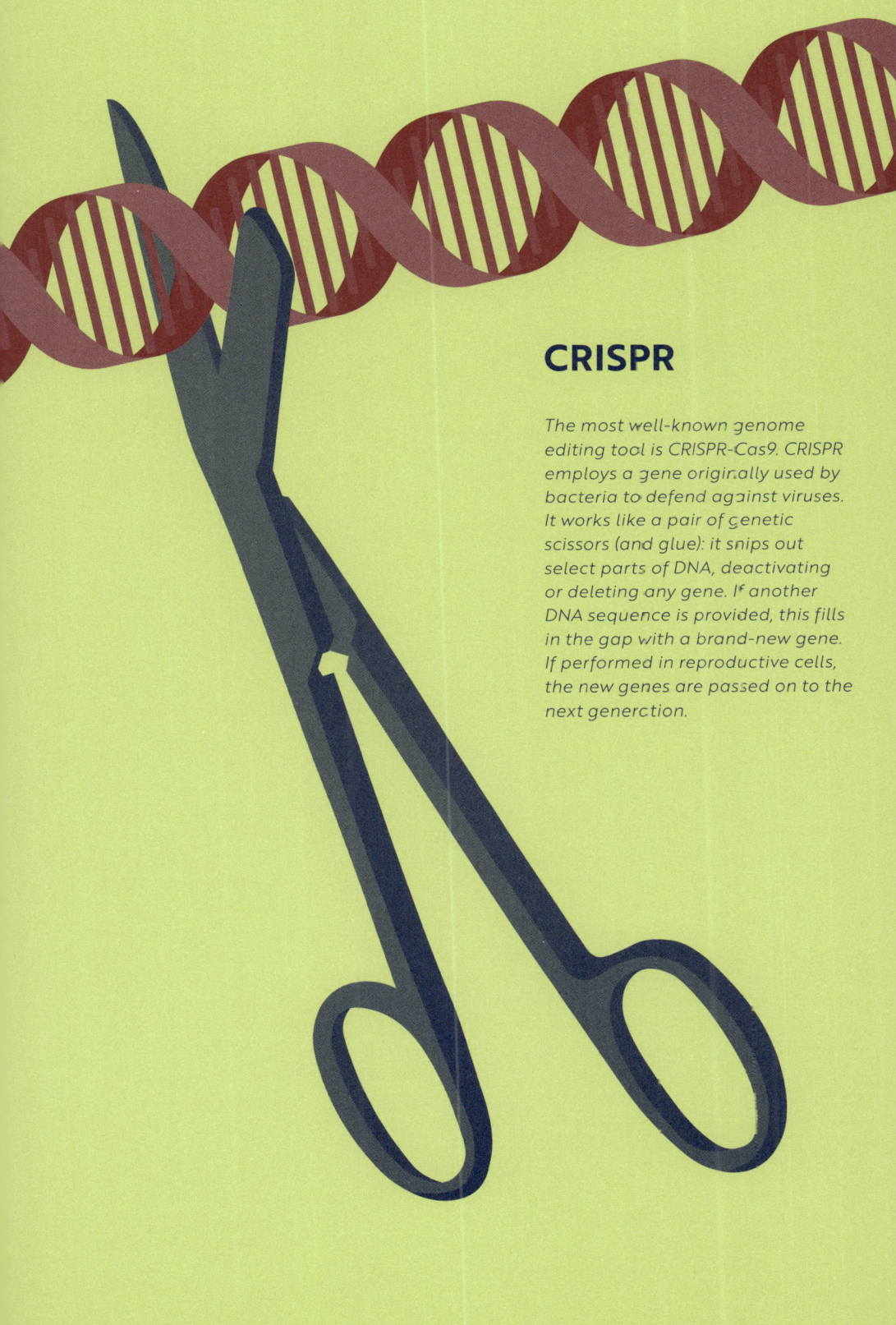

CRISPR

The most well-known genome editing tool is CRISPR-Cas9. CRISPR employs a gene originally used by bacteria to defend against viruses. It works like a pair of genetic scissors (and glue): it snips out select parts of DNA, deactivating or deleting any gene. If another DNA sequence is provided, this fills in the gap with a brand-new gene. If performed in reproductive cells, the new genes are passed on to the next generation.

Can a river have rights?

→ **One in New Zealand already does. Conservation through legal means has delivered successful protection for biodiversity and the environment. International agreements can effect change but do have weaknesses, as everyone has to agree and then duly implement them.**

When the moratorium on commercial whaling was enacted in 1986, it came in the wake of nearly three million killed whales during the 20th century. Despite some strong opposition, the ban passed and almost certainly saved several whale species from extinction.

International agreements can bring significant gains in conservation, but aren't perfect. The Convention on International Trade of Endangered Species (CITES), signed in 1973, was designed to protect and ensure the survival of threatened animals and plants by prohibiting their trade on international markets.

CITES represents a significant global move to protect biodiversity. Over 80 species have moved to a lower threat category, suggesting these protections may have worked. Yet CITES has critics. Decisions over trade and sanctions aren't free from politics, and membership is voluntary. CITES also only regulates legal trade, and illegal trafficking is still rife: between 2010 and 2018 over $2.4 billion was spent tackling it. Species can face long delays – a decade on average – between being put on the IUCN Red List and being protected under CITES.

In 2022, the UN adopted the Kunming-Montreal Global Biodiversity Framework, an international accord to protect the world's biodiversity. It aims to halt and reverse biodiversity loss, with specific goals including the protection of 30 per cent of the world's land and sea by 2030. The agreement is promising, yet has faced early criticism around funding and isn't legally binding. The year 2023 saw the adoption of a landmark treaty over the world's international waters, which if ratified would set up a framework to establish protected areas in the high seas. Implementation could be complicated and lengthy.

Nations also act alone. In 2017, New Zealand passed a law granting the Whanganui River human rights, meaning it can be represented in court. National and international agreements can succeed in protecting biodiversity but ultimately depend on the sharpness of their legal teeth. Three countries still hunt whales: nearly 40,000 of them since 1986.

SENTIENT ANIMALS

In recent years, recognition of animal sentience in national and international law has grown. These laws support the notion that certain non-human organisms can feel things like pain, boredom or joy – and perhaps even have consciousness, as scientists now think that other animals like cephalopods and birds have the neurological faculties to generate it. Ethical considerations about animal suffering are feeding into discussions around rights. Industrial octopus farms are on the horizon. Should they be?

What role can indigenous peoples play in conservation?

⟶ **Perhaps it's best to look at the past first. Indigenous peoples have lived sustainably within many parts of the world considered wilderness, and have shaped and protected biodiversity, for tens of thousands of years.**

For decades, a dominant principle for the protection of biodiversity suggested natural ecosystems will return to a more resilient state in isolation from humans. This dualistic model is known as 'fortress conservation' and separates humans from nature. It has come at the expense of many local peoples who have been evicted to create protected areas.

There have been growing calls to revise this model – which is still popular in some countries – moving to one in which indigenous peoples are instead supported to conserve the land. This is key, because around a third of people living in tropical countries depend on nature for their basic needs. There is mounting evidence from around the world that rather than hindering biodiversity, traditional management has fostered it.

Maps of 'wilderness' areas and global biodiversity hotspots overlap with diverse communities of humans who have co-evolved with nature for millennia, such as in the tropical forests of South-east Asia and New Guinea.

In Australia, data suggest that indigenous peoples have shifted tree composition to sustainably procure food, and by using 'cool' low-intensity fires have cleared grassland, boosted plant biodiversity, supported keystone species and prevented wildfires such as those ravaging the country today – for up to 65,000 years. In Canada, indigenous forest gardens have more diverse plant life than non-managed forests. Evidence suggests human intervention in the Amazon has boosted biodiversity, and through an ancient agricultural innovation created *terra preta*, a nutrient-rich 'anthrosol' – human-modified soil. Soil scientists are now studying its potential for biodiversity conservation.

Areas under indigenous and traditional management, recently coined 'Territories of Life', are some of the most biodiverse on the planet. They cover around 21 per cent of land on Earth and a third of intact forests.

A new conservation vision is emerging based on protecting the rights of indigenous and local peoples and pushing for conservation initiatives, which supports relations between nature and humans and promotes biodiversity in tandem.

SACRED CONSERVATION

In recent years, conservationists have grown to realise the extraordinary benefits sacred spaces can bring to biodiversity. There could be hundreds of thousands of wooded areas in countries across disparate continents, all protected through their designation as an important religious site. These groves or forests tend to have far higher biodiversity than their surrounding environments, and offer shelter to many endangered species. Sacred areas are natural genetic banks for flora and fauna, particularly threatened and endemic species.

What can I do to protect biodiversity?

→ **Human behaviour is linked with both the destruction and preservation of biodiversity in our age. We can all help by monitoring biodiversity, supporting change and altering our consumption patterns.**

Everyone can contribute to the protection and preservation of biodiversity. By volunteering to help monitor life and the environment, people can engage in citizen science. This practice has grown in popularity over recent decades, and is increasingly recognised as a valuable contribution to scientific research. Citizen science can involve the collection of data, such as environmental variables and weather patterns, or by reporting sightings of animals.

This kind of monitoring can be incredibly useful to scientists, especially in parts of the world where resources are stretched or remote sensing cannot reach. Established networks exist in the European Union, North America, New Zealand and Australia and are emerging in Africa, Asia and South America. As smartphone and mobile technology penetrates deeper around the world, participation is growing too.

Shifting narratives around conservation are highlighting the distorted yet widespread link between population growth – specifically the notion of 'overpopulation' – with environmental degradation and biodiversity decline. Rather, it is the way in which humans live that has a greater effect, particularly in regard to consumption patterns.

If unsustainable and excessive consumption is the main or at least one of the main drivers of global diversity decline, then this offers a way for every individual to effect change. Lowering meat intake (particularly beef), using less energy and water, and consuming fewer products will all reduce pressure on resources, land and natural habitats. A vegan diet cuts someone's impact on biodiversity by 65 per cent – and reduces greenhouse gas emissions to a quarter. What you eat really makes a difference. Supporting political action towards protecting biodiversity can help, along with those institutions working to save or promote it. Reducing waste and chemical use will lower pollution.

Stemming today's biodiversity decline will take collective and persistent changes in human behaviour on a global scale. Much of this will be driven by national and international governments. Yet every individual can play their part too. Over time, humanity can regain its collective connection to the natural world and work to protect the biodiversity that offers us so much – and like us, has a fundamental right to exist in and of itself.

THE VALUE OF LIFE

For decades philosophers have argued that nature should be valued for its intrinsic worth – not simply for the benefits we gain from it. These arguments are rooted in moral and ethical principles towards nature. Humans have an inherent fundamental need to protect things other than themsevles, they suggest, one which can be met through the protection of the natural world. This replaces an anthropocentric view of our role in nature with an ecological or 'life-centred' one. Our concern for other species and ecosystems will ultimately lead us towards more fulfilling lives.

FURTHER EXPLORATION

BOOKS

Antonelli, Alexandre. *The Hidden Universe: Adventures in Biodiversity.* Ebury Press, 2022

Carson, Rachel. *Silent Spring.* Houghton Mifflin, 1962.

Clover, Charles. *Rewilding the Sea: How to Save Our Oceans.* Witness Books, 2023

Kolbert, Elizabeth. *The Sixth Extinction: An Unnatural History.* Bloomsbury Publishing, 2014

Montgomery, David R. *Dirt: The Erosion of Civilisations.* University of California Press, 2007

Wilson, Edward O. *The Diversity of Life.* Penguin, 2001

Yong, Ed. *I Contain Multitudes: The Microbes Within Us and a Grander View of Life.* Ecco Press, 2016

ONLINE RESOURCES

Convention on Biological Diversity
www.cbd.int

The Guardian biodiversity coverage
www.theguardian.com/environment/biodiversity

Inkcap Journal: Longform journalism about nature and conservation in Britain
www.inkcapjournal.co.uk

IPBES Global Assessment on Biodiversity and Ecosystem Services 2019
www.ipbes.net/global-assessment

Nature Portfolio: Research and news on biodiversity
www.nature.com/subjects/biodiversity

PODCASTS

Babbage, The Economist
play.acast.com/s/theeconomistbabbage

BBC Inside Science
www.bbc.co.uk/programmes/b036f7w2

BBC Naturebang
www.bbc.co.uk/programmes/m00060x0

The Case for Conservation Podcast
www.case4conservation.com

Nature Insight: Speed Dating with the Future (IPBES Podcast)
podcasts.apple.com/us/podcast/nature-insight-speed-dating-with-the-future/id1518308737

Radiolab
radiolab.org

The Wild with Chris Morgan
www.kuow.org/podcasts/thewild

NOTES ON CONTRIBUTORS

AUTHOR

Richard Kemeny

Richard is a writer specialising in science and the environment, with a particular emphasis on climate change, ecology and biodiversity. His work has featured in *National Geographic*, *BBC Travel*, *Smithsonian*, *New Scientist*, *The Economist*, *Hakai Magazine*, *The Atlantic*, *Wired UK*, *MIT Technology Review*, *Science* and *Sapiens*. He has also done radio and podcast work for PRI's The World, the BBC World Service and Economist radio, and been awarded fellowships at the Woods Hole Oceanographic Institution, the Marine Biological Laboratory and the Erice International School of Science Journalism.

ILLUSTRATOR

Robert Fiszer

An illustrator and designer based in London, Robert creates conceptual vector illustrations for books, web and advertising. His style is clean and minimalist, ensuring that the viewer is drawn directly to the essence of the image and its meaning. www.robertfiszer.com

INDEX

A
abiotic 30, 32, 34, 43
acoustic lures 89, 94
agriculture 58, 71, 76, 77, 108, 112, 125, 126
agroforestry 71, 77
algae 50, 53, 56, 62, 124, 132, 133
Amazon rainforest 26, 108, 152
Anaxagoras 50, 52
animals 38, 92–3, 100, 104, 110, 111
   domestication of 107, 112
   feralisation of 107, 113
   sentient animals 151
Antarctica 56, 92, 116
anthrome 105, 106
the Anthropocene 11, 102–19
antibiotics 54, 71, 80
antifreeze gene 16
antimicrobial resistance 71, 78
archaea 50, 51, 53, 54, 57, 58
Arctic 26, 94, 100, 116, 147
Aristotle 18, 19
Arrhenius, Svante 50
artificial intelligence (AI) 90, 93, 100–1
Asahi, Mount 56
aspirin 80
atmosphere 10–11, 62, 69

B
bacteria 50, 51, 53, 54, 57, 58, 62, 71, 78, 133, 149
Bartsch, Paul 92
Becking, Baas 51
bees 98, 132
binomial naming system 18, 19
biobanks 141
biodiversity: biodiversity hypothesis 78
   data collection 86–101, 154
   decline in 31, 76, 90, 100, 105, 118, 122, 123, 126, 128, 130, 145, 146, 154
   definition of 13
   hotspots 13, 26–7, 98, 152
   valuing 22–3
bioenergy 70, 74
biological carbon pump 32, 39
biomes 33, 34
biosecurity 125
biospheres 13, 24, 86
biotic 30, 33, 34
biotic homogenisation 107, 118, 123
birds 92, 93, 94, 110
bison 16
breeding programmes 138, 141, 142
Buchner, Johann Andreas 80
bushmeat 125, 128
butterflies 42–3
by-catch 125, 128

C
camera traps 87, 88
CAPTAIN 100
carbon 26, 30, 38–9, 50, 53, 56, 58, 70
   carbon cycle 32, 38, 60
   carbon sinks 32, 38, 58
   carbon systems 32–3

reducing emissions 105, 139
   sequestration of 69, 71, 72, 146
carbon dioxide 11, 32, 35, 38, 118
carcasses 36–7, 39, 40
cats, feral 135
Ciamician, Giacomo 75
cities 110–11
citizen science 101, 154
classification systems 18–19
climate 22, 69, 100, 105, 107
climate change 11, 14, 20, 24, 31, 58, 108, 123, 130
   biodiversity decline 145
   hotspots and 26
   and invasive species 134
   mangroves and 72
   mitigating 147
   ocean microbes and 62
   and species conflict 42, 43
coastal ecosystems 73
commensalism 32, 44
competition 33, 42–3, 46, 131
conservation 22, 82, 86, 90, 95, 100, 136–55
consumption 36, 126, 154
Convention on Biological Diversity (1993) 139
Convention on International Trade of Endangered Species (CITES) 150
coral and coral reefs 22–3, 44, 73, 90, 94, 98, 100, 116, 118, 130, 148
crabs 44, 116
CRISPR-Cas9 149
crops 76, 77, 107, 112, 125, 126, 144
CryoArks 141, 144

D
Daily, Gretchen 69
Darwin, Charles 12, 16, 43
Darwinian demons 47
dead zones 125, 133
deep ocean 26, 32, 38, 56, 116
diel vertical migration 39
diseases 20, 78, 79, 80, 125, 134, 144
diversity, origins of 8–27
DNA (deoxyribonucleic acid) 12, 16, 17, 97, 141, 147, 149
   environmental DNA (EDNA) 88, 96–7, 98
domestication 107, 112
doppler sensors 89, 98
drones 90, 98
Dust Bowl 58–9

E
EarthBiogenome Project 145
earthquakes 24, 93
ecological doom-loops 123, 124
ecological succession 32, 46
ecosystems 18, 20–1, 28–47, 78, 94, 110
   climate change and 130
   collapse of 123
   ecosystem services 68–9, 70, 72
   holobiont 52, 55
   hotspots 26
   humans' impact on 126
   indicators of health of 37, 44, 60, 90
   nested ecosystems 63
ecotourism 70, 72, 82–3
Ehrlich, Paul and Anne 68

elephants 42, 43, 87, 89, 91, 140
energies, renewable 70, 74
environmental DNA (EDNA) 88, 96–7, 98
essential biodiversity variables (EBVS) 86, 89
eukaryotes 10, 13, 53, 57
extinction 42, 113, 122, 124, 131, 134
   background extinction rate 107
   de-extinction 140, 147
   mass extinctions 14, 118–19
   Quaternary Extinction 124, 128
extirpation 122, 125
extremophiles 53, 56, 57, 80

F
faecal transplants 54
feralisation 107, 113
fish 16, 40, 44–5, 72, 96, 100, 116, 128, 134
food 126, 154
   food chains 33, 36
   food web 33, 36, 40, 53, 56, 62
forests 32, 38, 69, 72–3, 126, 127, 152, 153
fortress conservation 152
fossil fuels 38, 71, 74, 104, 105, 108
fossils 105, 106, 109, 118, 119
frogs 96
fuel 74
fungi 50, 51, 52, 54, 58, 60–1

G
Gaia Hypothesis 11, 13
garbage patches 106, 117
genes 13, 16, 17, 141, 148–9
genetics 12–13
   diversity 18, 20, 64, 141, 144
   genetic engineering 80, 139, 141
genomes 64, 141, 145
genotypes 141
geologic time scale 104, 107
geological processes 24
geophones 87, 89
Global Biodiversity Information Facility (GBIF) 87
Global Positioning Systems (GPS) 89, 98
goats, mountain 42
Golden Spikes 106, 109
gorillas 82
the Great Acceleration 108
Great Atlantic Sargassum Belt 132
Great Pacific Garbage Patch 117
greenhouse gases 108, 154
Gross Domestic Product (GDP) 69, 70
Gross Ecosystem Product (GEP) 69, 70
gut microbiome 54, 111, 142
gyres 106, 117

H
Half-Earth resolution 139
Hardin, Garrett 129
health 78–9
holobiont 52, 55, 63
Holocene 104, 106
homogecene 123, 125
hotspots, biodiversity 26–7, 98, 152
humans 44, 54–5, 64, 97, 104, 108, 112, 114
hybridisation 107, 123
hydrophones 87, 89

hydropower 70, 74
hydrothermal vents 35, 53, 56, 62

I
inbreeding depression 141
indigenous peoples 152
intercropping 125, 126
International Commission on Stratigraphy (ICS) 105
International Union for the Conservation of Nature (IUCN) 138, 142, 143, 150
Internet of Animals (IOA) 88, 93

J
jellyfish 98, 148

K
Kunming-Montreal Global Biodiversity Framework 150

L
Leeuwenhoek, Antoine van 50
LiDAR technology 89, 90
Lindeman, Raymond 30, 33
Linnaean System 18
Linnaeus, Carolus 12, 18
lions 118
Lovelock, James 11

M
Man and the Biosphere Programme 31
mangroves 71, 72-3
Margulis, Lynn 11
Max Planck Institute 98
medicines 56, 76, 80-1
megafauna 140, 146
mental health, biodiversity and 78
microalgae 53
microbes 14, 50-65, 71, 74, 78
microbiome 51, 52, 54, 58, 59, 111
microbiota 50, 53, 142
microscopes 50
migration 106, 114, 115, 130
Milankovitch Cycles 105, 107
monoculture 71, 77, 125
mosquitoes 148
mountains 13, 24-5
multicellular organisms 10, 13
mutualism 40, 44, 60
mycorrhizae 51, 52, 60-1
Myers, Norman 26

N
national parks 138, 140
nature reserves 138, 140

O
oceans 10, 11, 14, 24, 64, 108
  acidification of 118, 130
  carbon and 38, 39
  deep ocean 26, 32, 38, 56, 116
  ocean microbes 62-3
  salinity 34-5

Odum, Eugene 31
One Health 79
Ostrom, Elinor 129

P
pangolins 128
panspermia hypothesis 50, 52
panthers, Florida 126
parasitism 32, 44, 134
Payne, Roger 95
permafrost 32, 38
phoresy 45
photosynthesis 30, 32, 36, 38, 44, 70, 74, 75
phytoplankton 36, 39, 53, 62
plants 38-9, 62, 74, 94, 104, 114, 122
plastics 62, 105, 106, 116-17
Plato 68
platypus 14-15
pollination 40, 76, 99, 112, 122, 132
pollution 58, 74, 78, 116, 126, 132
polychlorinated biphenyl (PCB) 124, 132
population 126, 154
prairie dogs 40
predators 33, 40, 42, 44-5, 46, 76
prokaryotes 10, 13, 51, 53
protected areas 83
protists 53, 54
protozoa 50, 53, 58
Przewalski's gazelle 142

Q
Quaternary Extinction 124, 128
quinine 80

R
rainforests 13, 20, 26, 90, 94, 100
regulating service 71, 72, 76
remora 44-5
remote sensing technology 88-9, 90-1
renewable energies 70, 74
rewilding 140, 146-7
rhizosphere 52, 59
rivers 34-5, 72, 150
RNA (ribonucleic acid) 10, 12
robotics 98-9

S
sacred conservation 153
salicin 80
satellite telemetry 86, 87, 88, 90, 91
scallops 40
seed banks 141, 144
sentient animals 151
sharks 40, 44-5, 92, 128
skin microbiome 54, 142
smelt 16
snow 56
snowball Earth 11, 13
soil 58, 59, 68, 69, 71, 76, 110, 116
solar power 70, 74, 75
sonar 87, 100
Soulé, Michael 138
soundscape ecology 87, 89, 94-5
species 12, 14, 20, 32, 46, 87, 122, 125
  diversity 18, 20, 34, 78, 80, 100
  endemic 24, 26, 27, 118, 130, 153

hybridisation 107, 123
identification of 18, 91
invasive 114, 122, 125, 134
IUCN lists 142, 143
keystone species 33, 40-1, 152
numbers of 51, 58, 60
saved species 142-3
speciation 12, 14-15, 16
species distribution models 86, 89
taxonomy 18-19
stingrays 40
succession, ecological 46
sustainable development 31, 139
Svalbard Global Seed Vault 144
symbiotic relationships 32, 44, 54, 55, 60, 62, 70, 83
synthetic biology 141, 148

T
Tansley, Sir Arthur 30
taxonomy 12, 18-19
technology 84-101
tectonic plates 24-5, 38
Territories of Life 152
therapeutics 80
thermoregulation 12
toads 135
tourism, ecotourism 70, 72, 82-3
tracking technology 92-3
Tragedy of the Commons 124, 129
translocation 140
trees 78, 98
trophic cascade 33, 40, 41
trophic dynamics 30, 33
trophic levels 33, 36
turtles 46-7, 92, 128

U
unicellular organisms 10, 13
unseen world 48-65

V
virosphere 53, 64, 65
viruses 50, 53, 54, 58, 64-5, 71, 78, 149
volcanoes 10, 13, 24, 38-9, 46, 56, 93
vortex, extinction 122, 124

W
Wallace, Alfred Russel 16
wasps 42, 76
weather 20, 58, 69, 72, 73, 122, 130
whales 36-7, 94, 95, 98, 150
wilderness 108-9, 152
wildlife trade, illegal 128, 150
Wilson, E.O. 18, 44, 139
Woese, Carl 57
wolves 41
Wood Wide Web 60-1

Y
Yellowstone National Park 41, 138, 146

Z
zoopharmacognosy 81

ACKNOWLEDGEMENTS

A huge thanks to my partner Sally Brown for your support, inspiration and for being a source of constant positive energy. I'm grateful for insightful and fun conversations with Sofia Castello y Tickell and Robert Smith about biodiversity, ethics and more. Special thanks to Michaela Agapiou for keen-eyed comments. Thank you to the scientists who took the time to chat about their work, and the many others whose research and debates formed the backbone of this book. I'm thankful for the editors at UniPress, particularly Katie Crous, for helping to shape it. And to the illustrator, Robert Fiszer, for bringing it to life.

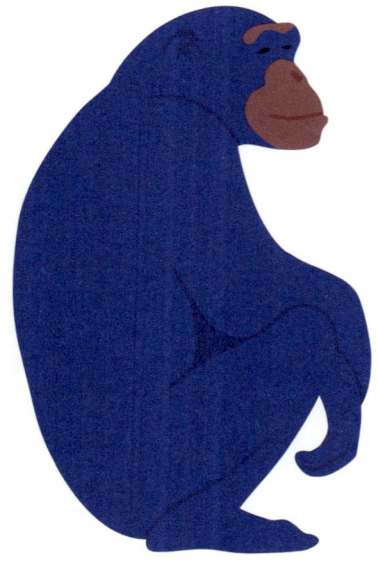